CW00689163

SN3 4PE

MINI 60 YEARS

CARNABY ST.
MARY QUANT
COLOUR SHOP
Designer
E477 LJO

MINI 60 YEARS

GILES CHAPMAN

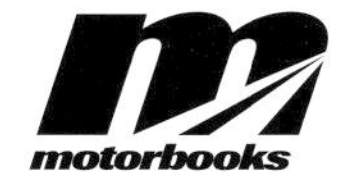

First published in 2019 by Motorbooks, an imprint of The Quarto Group, 100 Cummings Center, Suite 265D, Beverly, MA 01915 USA. T (978) 282-9590 F (978) 283-2742 www.QuartoKnows.com

Motorbooks titles are also available at discount for retail, wholesale, promotional, and bulk purchase. For details, contact the Special Sales Manager by email at specialsales@quarto.com or by mail at The Quarto Group, Attn: Special Sales Manager, 100 Cummings Center, Suite 265D, Beverly, MA 01915 USA. T (978) 282-9590 F (978) 283-2742

10 9 8 7 6 5 4 3

ISBN: 978-0-7603-6399-7

Digital edition published in 2019
eISBN: 978-0-7603-6400-0

Library of Congress Cataloging-in-Publication Data
Names: Chapman, Giles, author.

Title: Mini : 60 years / Giles Chapman.
Description: Minneapolis, Minnesota : Motorbooks, 2019. | Includes index.
Identifiers: LCCN 2018050937 | ISBN 9780760363997 (hardcover + jacket)
Subjects: LCSH: Mini-Cooper automobiles--History.
Classification: LCC TL215.M465 C47 2019 | DDC 629.222/2--dc23
LC record available at https://lccn.loc.gov/2018050937

Acquiring Editor: Darwin Holmstrom
Project Manager: Alyssa Bluhm
Art Director: Laura Drew
Cover Designer: James Kegley
Covers Illustrations: Ed Jackson
Layout: Amy Sly

Printed in China

DJB 92B
ORX
707F
1968 185 1968
RALLYE MONTE-CARLO

620 ERX

15 Successful Years
of the MINI 1959 - 1974

HOK 657W
HOK 658W
HOK 671W

TABLE OF CONTENTS

5,000,000
MINI
AUSTIN ROVER
5,000,000th
MINI
5,000,000 MINI
AUSTIN

INTRODUCTION

Throughout six decades, Minis have been some of the most charismatic cars you can buy. From a driving point of view, the experience has always been joyful—scintillating, even—whether the power on offer is meager or mighty. A lot of people *have* to own a small car, but almost everyone who actively enjoys getting behind the wheel *wants* to own a Mini.

Minis are also controversial.

The original car was the brainchild of a single individual, Alec Issigonis, who had the opportunity to reshape mass motoring like few before or since. He mixed the forcefulness of Henry Ford and André Citroën, the engineering standards of Ferdinand Porsche, and a boyish enthusiasm that P. G. Wodehouse would have been proud to create. His Mini was conceived in troubled times as a radical new minimum-standard for economy cars. Everything about it was daring.

The Mini went on, both by design and unexpected circumstance, to be a constant in Britain's motoring scene for an incredible forty-one years. The last car made in 2000 really was very little different from the first one in 1959. Its survival throughout that time is an eye-opening story in itself.

Replacing the Mini with the MINI proved to be one of the hardest assignments any car manufacturer could undertake. BMW grabbed the mettle and chose to reincarnate the hard-charging Mini Cooper for the new millennium. The Anglo-German process was fraught, but the result was a compact car that stoked desire and pride in a new way while offering every safety, comfort, and convenience feature the twenty-first century consumer expects as a matter of course. Totally updated twice more since 2001, the MINI has retained its distinctiveness while blossoming into an entire range of cars.

Trying to write an overview of both the Mini and the MINI was never going to be easy. What I hope I've achieved is an explanation of the why, the how, and the what of the two cars, as well as the way the why informs the how. I have friends and acquaintances who own both, and there's often tension because, although people who drive the modern cars have warm feelings for the originals, diehards of the 1959–2000 Mini rarely seem to like its successor. They feel it's too bulky, too fancy, and has taken liberties with the minimalism that Issigonis espoused. Little you can say will change their minds, and such passion must be respected, and rather admired.

But there is no doubt that the 2001-on MINIs are excellent, and not least for the high-quality standards and enormous opportunity for personalization they offer. In common with the rest of the automotive world, the MINI is now entering its electric phase and beginning to sideline the internal combustion engine, which was all Issigonis understood about powering a car. While this will be a very different motoring era to the one he knew, any driver who loves cars with character must hope the MINI continues to maintain its very characterful place within it.

In 1986, TV presenter and all-around car nut Noel Edmonds (standing at door) was given the honor of driving the five millionth Mini off the production line.

CHAPTER

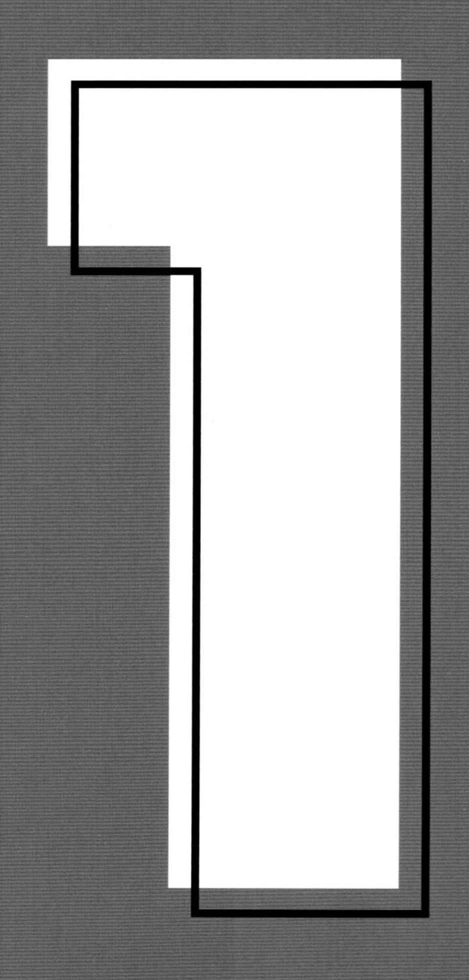

CARS FOR THE PEOPLE

1955-D

OUT OF CRISIS AND CONTROVERSY, THE PLANETS WERE SUDDENLY ALIGNED, AND THE ENVIRONMENT WAS RIPE FOR AN ECONOMICAL NEW VEHICLE THAT ANYONE COULD ASPIRE TO.

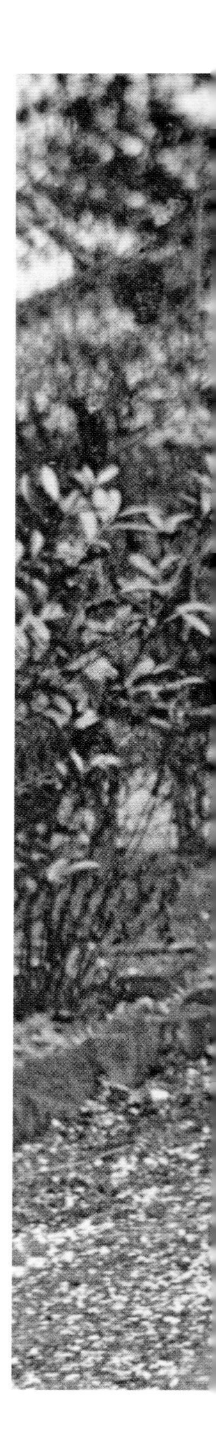

Long, long before any speck of the Mini was even detectable in the mid-1950s, the quest for a convincing "people's car" had propelled the motor car forward. Because, at first, it was little more than a volatile plaything for the ultrawealthy—a toy for Sunday-afternoon outings and, once engineers had learned how to harness power and roadholding, a mechanical racehorse to provide thrills for drivers and a spectacle for the crowds.

Making motoring an affordable right for every citizen was really the impetus behind German government thinking in the 1930s. The Volkswagen, or "people's car," was actually how the German Labor Front framed it in 1934. No matter how twisted and dastardly the regime's overall ethos, the Nazis' car offering was irresistible to the man in the street then navigating on foot, bicycle, or motorbike. The Kraft durch Freude car was part of a project of the same name (translating to "Strength through Joy") dedicated to making formerly middle-class pursuits such as holidays and travel available, and affordable, for everybody. Professor Ferdinand Porsche looked after the engineering of the rear-engined four-seater saloon, or cabriolet, designed to be "foolproof equipment." Meanwhile, the Third Reich invented a groundbreaking installment-plan scheme allowing even relatively poor people to buy a Volkswagen—a coupons-based savings plan through which Germans could order and pay for a KdF car at 990 reichsmarks in weekly bites of just 5 reichsmarks. By 1939, almost 170,000 savings accounts had been opened by people keen to get one.

The scheme, like the whole Nazi project, was ultimately doomed, although the Volkswagen model we all know now as the Beetle did go on to massive success.

Back in 1908, however, the private enterprise of Henry Ford had achieved something very similar. By introducing automation and mass production to the assembly lines of his Detroit plant, he was able to lower the price of his Model T—a rugged, simple car whose huge ground clearance made it ideal for America's largely unmade roads—so that even the men who built it could afford one. Sales absolutely exploded, and Ford sold fifteen million Model Ts over the next nineteen years.

The Ford Model T was perfectly conceived for America's rough roads, with high ground clearance and rugged construction. Its design, together with Henry Ford's super-efficient mass-manufacture techniques that drove down prices, led to multimillion unit sales.

[above] This small car from Citroën, introduced in 1922, was the Type C 5CV; it took account of consumer feedback and so featured an electric starter instead of a starting handle to appeal to the growing band of female drivers.

[right] In Britain, Herbert Austin astutely realized that what new drivers really wanted was a "normal" car only in economical, miniature form. The delightful Austin Seven was his answer in 1922.

The name means "Little Mouse," and sometimes Mickey Mouse, but Fiat's 1936 500 Topolino was real progress rather than a Disney fantasy—a compact city car that was stylish and easy to use.

In 1936, Fiat introduced its 500 Topolino (or Little Mouse), fully intending to bring a small, economical car to a market that was widening out from just the professional middle classes. It was a two-seater aimed at city life, but it was notable for boasting the scaled-down features of much larger cars, such as a water-cooled four-cylinder engine, streamlined styling, and a four-speed gearbox.

In the middle of the era between the Model T and the Beetle, the Topolino was Britain's highly significant contribution to the growth in mass car ownership. In 1922, Herbert Austin introduced the first "real car in miniature," the Austin Seven with four wheels, four seats, and four cylinders. The monied drivers of the Roaring Twenties joked that you'd need to buy two "Sevens," one for each foot. But the car's concept worked a treat. Its sophisticated design, with compact dimensions and low running costs—and a starting price of £122—was the death knell for most of the motorbike-derived "cyclecars." The Seven also went global, with licensed manufacture in Germany, France, and the United States, and was closely copied by Japan's Datsun. Citroën's Type C also reflected very similar thinking, but it strode one stage further into the arena of consumer products by offering an electric starter; this convenience, in place of a starting handle and the requisite brute force needed to crank it, hugely increased its appeal to women drivers.

Providing affordable motoring to the everyman dropped to a very low priority in 1945 in Europe, as the world emerged dazed, exhausted, and impoverished from the turmoil of the Second World War. All around was government-enforced austerity and rationing, while many car factories had been left as rubble-strewn wrecks after being pounded during the bombing raids of the conflict. Roads across Europe were in a parlous state, and gas was usually hard to come by and reserved for drivers who commanded priority, such as doctors.

The first sign of new products within reach of the ordinary driver came in 1947, when Renault revealed the 4CV with its startling Volkswagen overtones, both in shape and in the rear-mounted position of its engine. The following year, however, would reveal the intriguing divergence in thinking between different manufacturers on what constituted the ideal popular small car.

In France, Citroën's 2CV was an exercise in extreme minimalism and frugality. Its air-cooled twin-cylinder engine, front-wheel drive, and long travel suspension

were combined to make it nimble and reliable anywhere in rural France, while the flimsy body, tubular hammock-type seats and canvas top were all ways to pare manufacturing costs to the bone.

Britain's answer came in the form of the Morris Minor. It was much more radical than its sober specification initially suggested because although the ultimate pulling power from a side-valve engine was nothing special, its precise rack-and-pinion steering and torsion-bar front suspension made it feel eager and pin-sharp. Its designer, a young chap by the name of Alec Issigonis, had built his own lightweight racing car in the 1930s, and so he very well understood the safety benefits of a car that was easy to control at speed and difficult to unseat in poor conditions. The Minor was destined to become a trusted favorite of suburban Britain during the early 1950s, but throughout that whole decade its superior dynamics continued to set it apart from mediocre competitors.

To anyone who hoped that global geopolitical bust-ups were now firmly in the past, the 1950s were to offer some unpleasant surprises. First, in 1950, came the Korean War, the effects of which would plunge world trade back into a spin, with materials and commodities diverted to a sudden increase in defense spending. Frosty relations between America and the Soviet Union meant that the old war gave way to a new Cold War. And then, in 1956, came the so-called Suez Crisis, when Egypt grabbed control of the Suez Canal from its British colonial occupiers. This narrow shipping lane is crucial to world trade movements; just look at a world map to see what the alternative route is for, say, a tanker full of crude oil—thousands of extra miles taking perhaps weeks in the roughest of seas.

What the troubles in far-off places highlighted was that fuel supplies were at the mercy of world events. The more that people built their lives around car ownership, the more dependent they were on their vehicles. If gas supplies were hit, the personal and economic impact could be huge. So there was a genuine imperative in the 1950s to make a gallon of juice go as far as was possible without reverting to riding a two-wheeler.

[top left] At the wheel of the one thousandth example of the Volkswagen is Ivan Hirst, the British Army major who masterminded the rebuild of the VW factory and the start-up of manufacture for the car that became the Beetle.

[top right] This is designer/engineer Ferdinand Porsche with a near-definitive prototype of the Volkswagen, the German Nazi government's vision of an affordable car for the common man.

–o [left] The Citroën 2CV, sometimes called the Tin Snail, was launched in 1948 to offer the poorest in French society basic rural or urban transport. The 2CV's technical secrets are revealed in this cutaway drawing showing the front-wheel drive, long-travel torsion bar suspension, and an air-cooled twin-cylinder engine.

⚲ [below] Volkswagen's giant Wolfsburg plant at full swing in the mid-1950s. Soaring global demand for the economical, well-made, and robust little cars would eventually see over twenty-one million examples built right up to 2003.

Fortunately, quite a few individuals and firms had been tackling this conundrum directly for a good while. Chief among these was British engineer Lawrie Bond, whose three-wheeled Bond Minicar, introduced in 1948 and powered by a 122cc single-cylinder motorcycle engine, gave incredible potential fuel economy of 100 miles per gallon. Five years later, an Italian refrigerator manufacturer called Iso made a quantum leap over the super-basic Bond with its Isetta, an ingenious four-wheeler, still with a single-cylinder moped-type engine. It was so astonishingly short that it could

In 1948, the Morris Minor set new standards of excellence for small British family cars; it was both enjoyable and safe to drive, if slightly underpowered, while also being roomy and packed with character.

THE SUEZ CRISIS

Britain gained control of the Suez Canal in 1882, although it had opened twelve years earlier after having been built by slave labor. The canal transformed global shipping by linking the Red Sea to the Mediterranean, a shortcut that saved thousands of nautical miles on the route between Western Europe and East Asia—and not least for oil tankers.

By 1950, however, Britain's proprietorial dominance was challenged by King Farouk of Egypt, through whose country the canal flowed and who demanded that British troops leave the Canal Zone. A year later, moreover, the Iranians took entire control of their own oil industry, ousting their British partners, who'd been exploiting it since 1909. In November 1951, Britain mobilized troops around Suez after Farouk, propped up by the West, was toppled by nationalist military leader Colonel Gamal Abdel Nasser, and Egypt declared a state of emergency. Tensions after the coup would remain high until the summer of 1956 when, with the British military presence contracting, Nasser seized control of the Suez Canal, and with it the vital income from the charges levied on global shipping to use it. He quickly nationalized the canal and made it a key part of the new Arab Republic of Egypt.

An irate Britain, with French and Israeli help, jumped into action with a military assault, and the canal was obstructed in the battle—causing a painful fuel squeeze in Europe as oil supplies had to go around "the long way." However, there was global condemnation for Britain's arrogant actions and, crucially, no support from the United States; Britain's action ended in withdrawal in 1957, and the Suez Canal has been an Egyptian asset ever since. It was a humiliating but decisive end to Britain's major colonial ambitions. Out of this dismal affair, though, came the impetus for a new car that could operate on the very barest of resources . . .

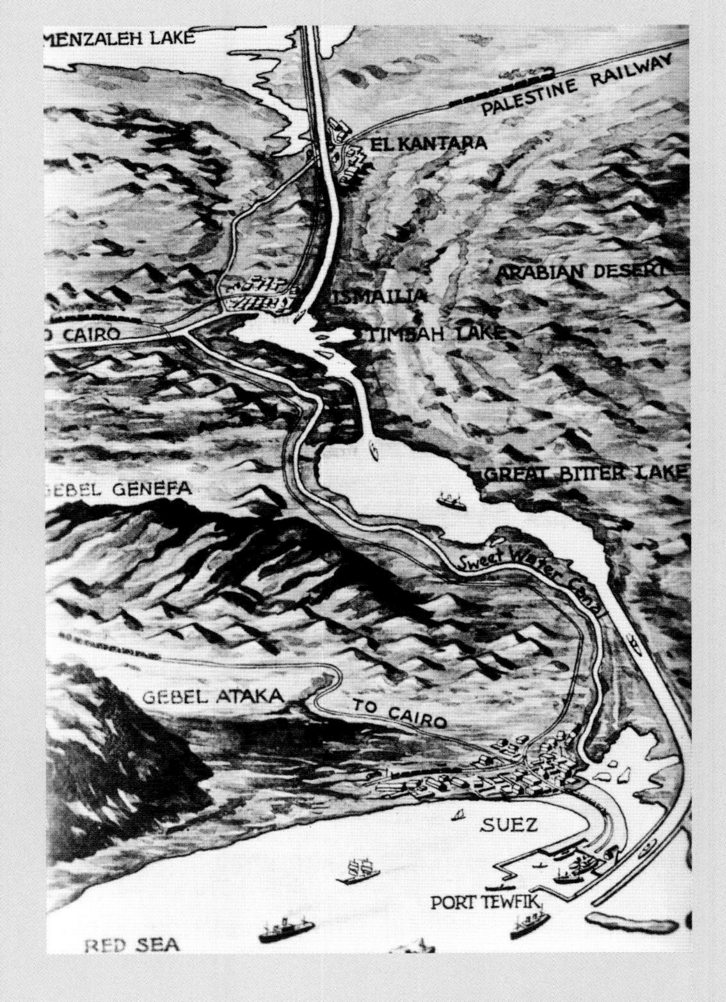

This delightful postcard image from the early twentieth century amply shows the strategic importance of the Suez Canal.

be parked end-on to the curb between other cars, and then the whole frontage swung open as a single door so the three occupants (mum, dad, and junior) could step right out on to the pavement. The Bond Minicar had had a simple canvas roof, but the Isetta was fully enclosed and rain resistant, and its egg-like appearance soon earned it a nickname: the "bubble car."

These tiny runabouts suffered mixed fortunes. They were enthusiastically adopted in Germany, where BMW bought exclusive rights to the Isetta and proceeded to sell 160,000 of them. There were homegrown rivals from former fighter-plane manufacturers Heinkel and Messerschmitt as well as the Fuldamobil Type S, Maico Champion, Zündapp Janus, and Goggomobil T and TS—all seeking to provide means of cheap personal transport that would snag consumer interest. These so-called microcars were somewhat less popular in France, Italy, and Spain, while in Britain—despite a plethora of imported and domestic offerings; the fact there was no road-tax concession on account of small dimensions or engine size, as there was in Germany, hampered take-up. Generally, only when designs were modified to become three-wheelers and could then be treated to the same low road-tax rates as motorbikes did they find much favor. These tiny cars tended

[top] Fiat's old 500 was refreshed and modernized in 1949 as the 500C, seen here in its Giardiniera station wagon version, and continued to be hugely popular throughout Italy and beyond.

[above] Through its very long life, and more than four million production total, the 2CV came to be considered a quirky choice for low-cost motoring, but its strengths of thrift, tenacity, and comfort remained undimmed.

The Bond Minicar was an interesting concept in ultraminimalism from Britain, benefitting from its classification as a motorbike-related tricycle to enjoy very low rates of road tax.

to be slow, feeble, unreliable, noisy, and manifestly unsafe and usually had smoky two-stroke engines, much as they could stretch out a gallon of gas over many miles.

Meanwhile, higher up the food chain, the Morris Minor was a good little car. It had been joined by other small saloons, such as the Austin A30 (the first from the company with integral body-hassis construction), the Ford 100E Anglia with its old-fashioned side-valve engine, and the cheap but dreary Standard 8. Britain's main chance of producing a true people's car had been passed over in 1945, when a delegation of British car-industry bigwigs traveled to Germany to examine the Volkswagen and its heavily damaged factory, with a view to taking it over as a spoil of war. Their report was damning and decisive: "This car does not fulfil the technical requirements, which must be expected from a motor car. Its performance and qualities have no attraction to the average buyer. It is too ugly and too noisy. Such a type of car can, if at all, only be popular for two or three years at the most." It's sobering to look at the subsequent success of the Volkswagen Beetle; it outsold the Minor almost nineteen to one.

Britain muddled on in its broadly uninspiring way, soon finding that the Volkswagen was providing formidable competition in the key US export market, where it set phenomenal new standards of reliability and toughness. Dealers loved selling it

because they rarely encountered an irate customer returning with a beef about the car.

Things car-wise might have continued in this complacent way indefinitely were it not for Suez (see sidebar on page 19). The panic it induced to conserve gasoline—even resulting in the issue of fuel ration books—focused every driver on how many miles to the gallon their cars could manage. Suddenly, potential car buyers were walking past the conventional offerings from the mainstream carmakers and looking again at the super-economical bubble cars with newfound keenness.

Minuscule cars like the Isetta suddenly posed a genuine threat to the established motor-industry order. High tariffs made imported cars uncompetitive, so instead there was a mini–gold rush by enterprising local businesses to secure manufacturing licenses on German designs like the Isetta, Heinkel Kabine, and Fuldamobil.

British manufacturers were alarmed, and at least one of them decided to mount a rear-guard action. Leonard Lord, the bombastic chief of the British Motor Corporation—formed in 1952 by the merger of the Austin Motor Company with the Nuffield Organization (see sidebar on page 23)—was typically forthright in his view.

"God damn these bloody awful bubble cars," he fumed at his technical-director colleague Alec Issigonis. "We must drive them off the road by designing a proper miniature car." Issigonis was just embarking on a project to replace BMC's midrange family cars, which were taking a beating from the Ford Consul and Zephyr in the sales war. That was now, all of a sudden, on hold.

[top] One of the more sophisticated of the 1950s German microcars, this Goggomobil TS300 offered minimal running costs with a touch of Alfa Romeo verve to its compact styling.

[above] Meet the first of the bubble cars, the BMW Isetta, with its single door consisting of the whole front panel hinged on one side and a single-cylinder motorbike engine at the back.

THE BRITISH MOTOR CORPORATION

When the makers of Austin and Morris cars pooled their resources in 1952, the resulting megacompany instantly became the world's third-largest car manufacturer. It was also the biggest exporter of cars to the United States, and its portfolio of car brands alone included Austin, Morris, MG, Riley, and Wolseley, in addition to vans, lorries, and tractors.

A merger had been considered several times before, and on paper at least it really was a merger—no shares, cash, or assets changed hands in either direction. However, it was clear from the start that forceful Austin managing director Leonard Lord would be dominant the new business and that Austin influence would play the dominant role in the creation of all future models.

There was a huge amount of overlap. Austin's A30 (shown above) and Morris's Minor were direct rivals, as were models like the Austin A50 Cambridge and Morris Cowley. Then at the end of 1952, the new Austin-Healey sports-car marque went head to head with MG. It wasn't until 1959 that rationalization really began to prevail, but even then, many cars were badge-engineered so that near-identical vehicles could be marketed through competing chains of Austin, Morris, Riley, and Wolseley dealers.

The infighting took years to subside, but in the interim BMC was under intense pressure from both the American-owned Ford and Vauxhall in Britain and hugely growing European marques, Volkswagen in particular, in world markets. There would be more mergers in the 1960s as BMC constantly found itself running just to stand still.

The forthright Leonard Lord (on right) examining the gas turbine "jet" engine in an Austin prototype; as head of the British Motor Corporation, he was determined to change the ethos of the small car.

CHAPTER

DESIGNING A CAR LIKE NO OTHER

ALEC ISSIGONIS AND HIS TEAM BROKE NEW GROUND ON MULTIPLE FRONTS AS THEY POURED THEIR ENERGIES INTO CREATING THE CAR . . . ADO15. HERE'S HOW.

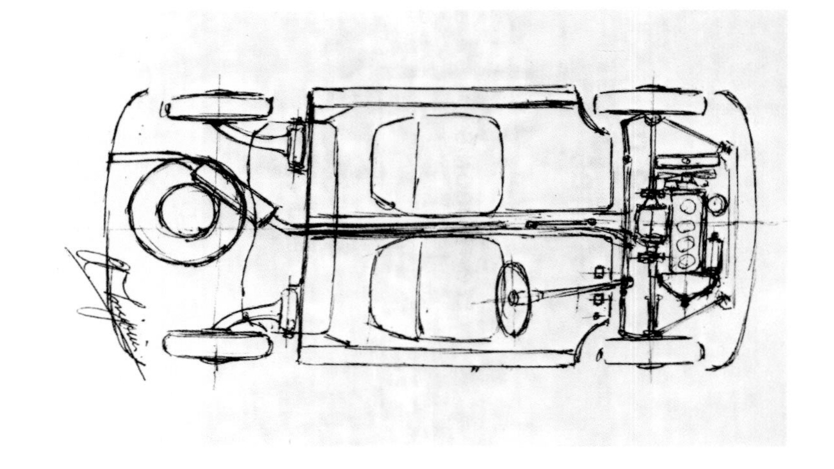

Having been charged with producing a completely new small car, Alec Issigonis had two compelling things in his armory of inspiration. One was a working Morris Minor prototype with front-wheel drive, which had itself been inspired by the pace-setting Citroën Traction Avant. It had been completed in 1952, during Issigonis's sojourn at Alvis, and was in regular daily use by very enthusiastic BMC engineers. The other was a ten-point plan that Issigonis had drafted for BMC chief Leonard Lord, setting out the hardpoints for the new car.

These two were presented in response to Lord's somewhat sketchy brief for a new small car. He had stipulated that it must be even more compact than the Minor, and extremely economical, yet still have enough interior space for four adult occupants. In other words, a proper motor car, unlike the feeble bubble cars, but one that would be sold at a rock-bottom price to convince every right-minded consumer to snub the Isetta and Bond Minicar. There was one more caveat: there could be no tooling up for a new power unit, and so therefore the car would have to use an existing engine.

This last restriction wasn't actually a problem for Issigonis. Although in the past he had lobbied passionately for innovative engines, such as a flat four for the Minor, this time he had just what he needed in BMC's four-cylinder A-Series. It was already a known quantity in the Minor and Austin A35.

Among Lord's demands, the final order was the toughest of all: there would be just over two years to design and develop the car and get it into production. This was in March 1957. Issigonis, however, was rubbing his hands in glee at the mammoth task ahead because he had negotiated total control over the entire project.

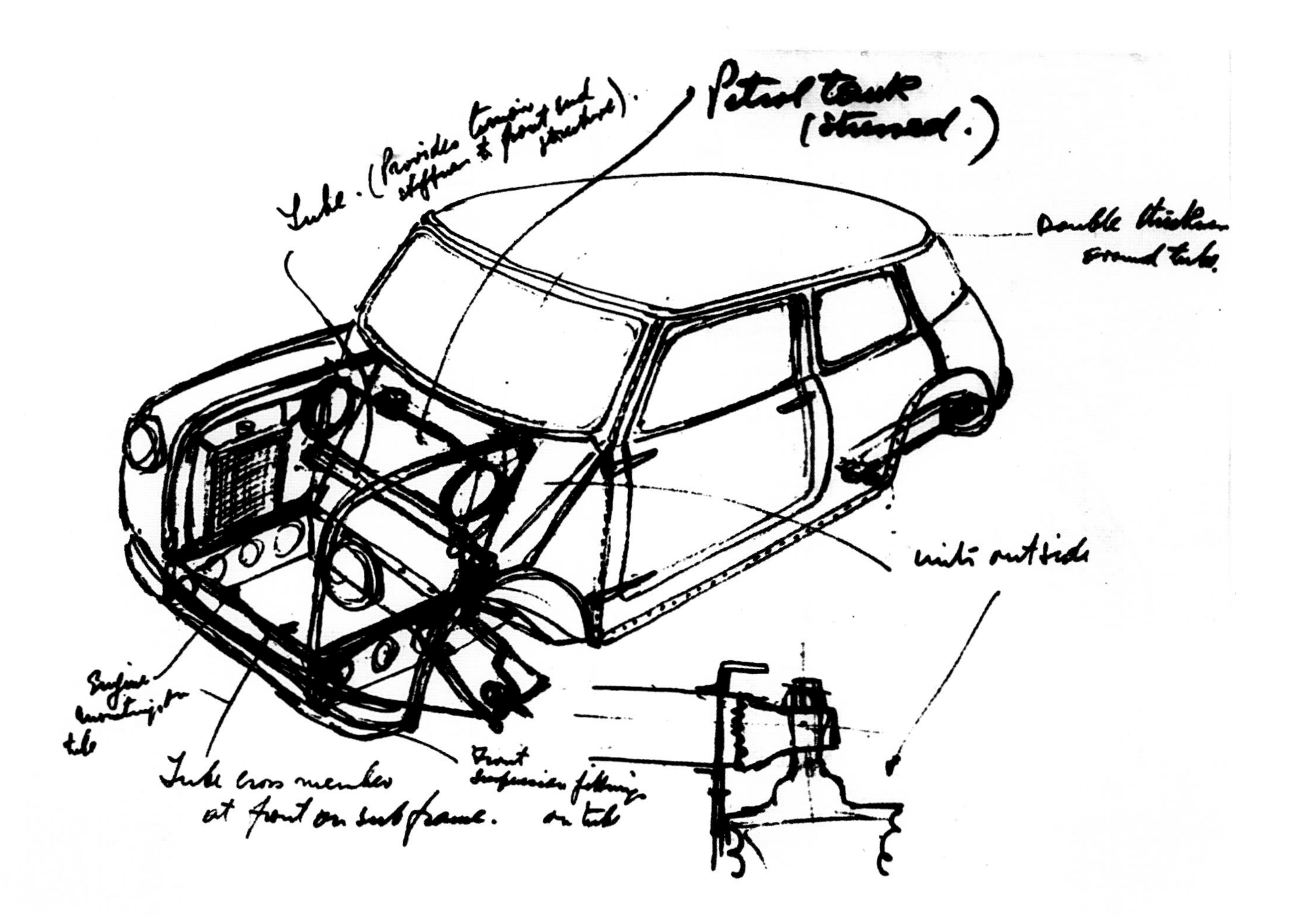

[above] Issigonis's famous sketch, rendered in 1958, of how he saw the car's structure, although at this preliminary stage a front subframe is only just becoming part of the deal.

[opposite] Another of Issigonis's pen portraits for the new economy car clearly shows the exhaust pipe running down its center channel, a key element of the car's torsional rigidity.

The excellence of the front-wheel-drive Minor put any thoughts of a rear-mounted engine out of his mind. In fact, the compact nature of the power pack in that car, in a spacious and lofty engine bay, helped frame his thinking about mounting the engine transversely—across the car east–west, rather than longitudinally, north–south—in the ADO15. An inveterate sketcher on paper, napkins, the proverbial cigarette packet, and even whole tablecloths, Issigonis in his early visualizations added a transverse engine position to a structure that pushed the wheels out to the very corners of a two-box structure: one box to house the engine at the front and another, bigger box behind to form the passenger compartment. Indeed, he could also draw upon work already done on a prototype called XC9001. This putative replacement for the Austin A55 Cambridge had the appearance of a huge four-door version of what would be the final Mini and was itself later developed into ADO 16.

There had already been a front-wheel-drive car with a transverse engine as far back as the 1930s. But the DKW Front of 1931 had a tall, narrow, twin-cylinder engine, which left enough room for its gearbox to be mounted alongside it. For BMC's new baby, the inline four-cylinder A-Series was too long for that when mounted across the car, filling all the space between the front wheels. This led Issigonis to conceive an innovative position for the gearbox: below the engine and inside the oil sump—or, rather, under the crankshaft in what became an extended oil sump, with the final drive unit behind it.

He had already decided that the car's length would be no more than 10 feet, and so naturally the packaging of the whole vehicle came in for close attention. Its footprint devoted 80 percent to passenger and cargo space. As there was no separate trunk volume overhanging at the back, this would naturally make for limited luggage accommodation, but Issigonis planned to use every last bit of space inside the car for stowage.

Early prototypes had difficulty coping with the strains put on their monocoque structure by the front-wheel-drive drivetrain, so the decision was taken to mount the main components in separate subframes front and back. Power was transmitted to the front wheels via constant-velocity joints, of a type created by Czech engineer Hans Rzeppa in the 1920s and adapted from a ball joint already in use in submarine conning towers. They consisted of a ball bearing surrounded by three cages, two of which were connected with the incoming and outgoing driveshafts, and a simpler flexible coupling for the inner end. This, in turn, allowed a sufficient steering angle without distortion or undue articulation, minimizing kickback in the steering.

The suspension was primarily a space-saving system, with minimal intrusion into cabin space, but was also highly effective. Issigonis had experimented with rubber-based suspension on his Lightweight Special in the 1930s, but this new system was perfected by his friend, the consultant engineer Dr. Alex Moulton. It used two compact metal cones with a layer of rubber in between instead of the usual coil, torsion, or leaf springs. The upper cone was bolted securely to the subframe; the lower rested on the wheel mount. As the rubber hardened under increasing pressure, this gave a

[bottom left] This 1957 styling mockup—note the lack of seats—shows the first thoughts on the front end, bumpers included, that would become familiar on the final Mini.

[bottom right] Not that you can tell from this rare archive image, but XC9003 was painted orange with a cream waist stripe and a black roof, leading to its nickname of the Orange Box.

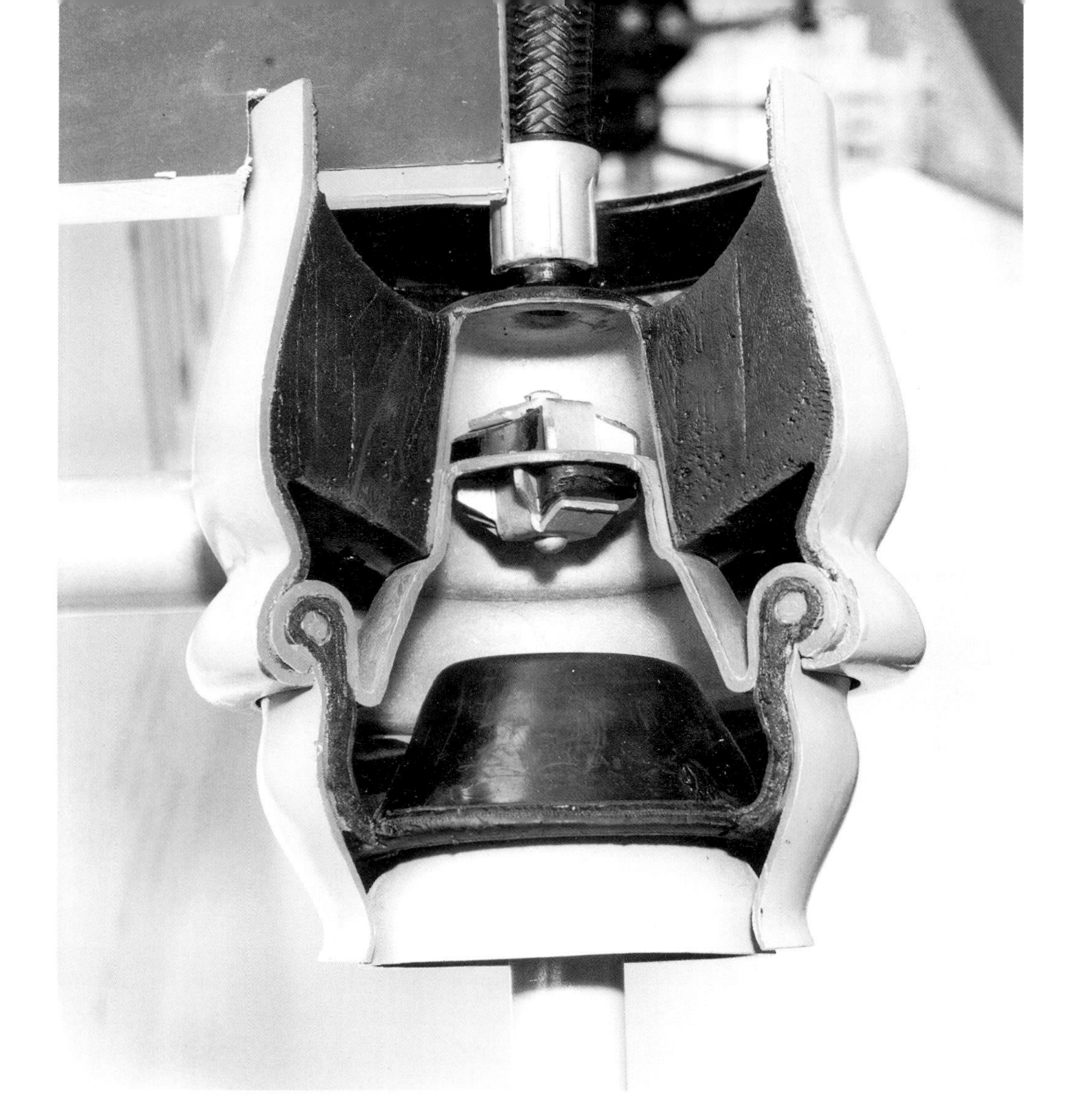

[left] A revealing view of one of Alex Moulton's rubber cone suspension units, a revolutionary system that gave the little car excellent handling, no matter what the laden weight.

[below] The entire Mini "powerpack" cradled by its front subframe and clearly showing the constant-velocity joint assemblies that took power to the two front wheels, and the long gear lever sprouting from the sump.

progressive suspension set-up well able to adjust from driver-only to four-up weight differences in the car. The lightweight assembly was so good at soaking up knocks that only small telescopic dampers—shock absorbers—were needed; they were fastened outside on upper front wishbones and rear longitudinal control arms for a smooth response to sudden pressures. Another measure intended to save space was 10-inch-diameter wheels, the smallest ever fitted to a conventional four-seater saloon, specially made by Dunlop to Issigonis's specification.

Space efficiency was an Issigonis obsession, and so it was key to the interior design as well. Using sliding windows not only contained manufacturing costs but freed up space in the doorframes that was given over to elbow room and deep storage bins for maps or handbags, molded into the trim panels. There were matching bins alongside the rear seats. A similar desire to open up storage space

CONTINUED ON PAGE 32

SIR ALEC ISSIGONIS

Alexander Arnold Constantine Issigonis, who went by Alec, is the father of the Mini and therefore an extraordinarily visionary character. Many men have designed and built a car to their precise liking; none have subsequently been as successful or influential as the Mini.

The only son of an Anglo-Greek engineer father and a German mother, he was born on November 18, 1906, in the Turkish town of İzmir (then called Smyrna, and once part of Greece). His father died in 1924, and he and his mother settled in England. Two years later, he took to the road in a small Singer car, driving his mother all over Europe in it and learning, en route, how to keep it running. This experience helped steer him towards a three-year mechanical engineering course at London's Battersea Polytechnic, although he was something of an unwilling student and was none too keen on math, which meant he left with a diploma instead of a degree. However, he showed he was an intuitive and gifted automotive engineer, preparing and racing his own intelligently modified Austin Seven Ulster, and he found no problem landing a job in the industry—starting in 1928 with Reduction Gears, where he worked on designs for a semiautomatic gearbox. He joined Humber in Coventry in 1934 to develop independent suspension systems, the prevailing new automotive technology of the era. Two years later he moved to Morris in Oxford to do the same thing and impressed his peers by building a single-seater racing car, his so-called Lightweight Special, which dripped with clever technical solutions and ingenious ways to cut excess bulk and was entirely handmade.

During the Second World War, Issigonis was closely involved in experimental military vehicles, but when time allowed he was able to refine plans for the Mosquito, Morris's crucial new car that would be launched as the Minor in 1948. It showed the astonishing scope of Issigonis's capabilities, and in that year he was appointed the company's chief engineer. By 1959 the Minor had proven so popular that it became the first British car to reach one million sales.

Unsettled by the Austin–Morris merger in 1952, in which he could see his influence waning, Issigonis accepted the post of technical director at Alvis, with a brief to develop an all-new V-8 luxury saloon. The total control over the project undoubtedly suited Issigonis's ego. Since the success of the Minor, he'd gotten quite used to, and gradually insistent upon, having things done his way or not at all. However, the three years of work he poured into the Alvis came to

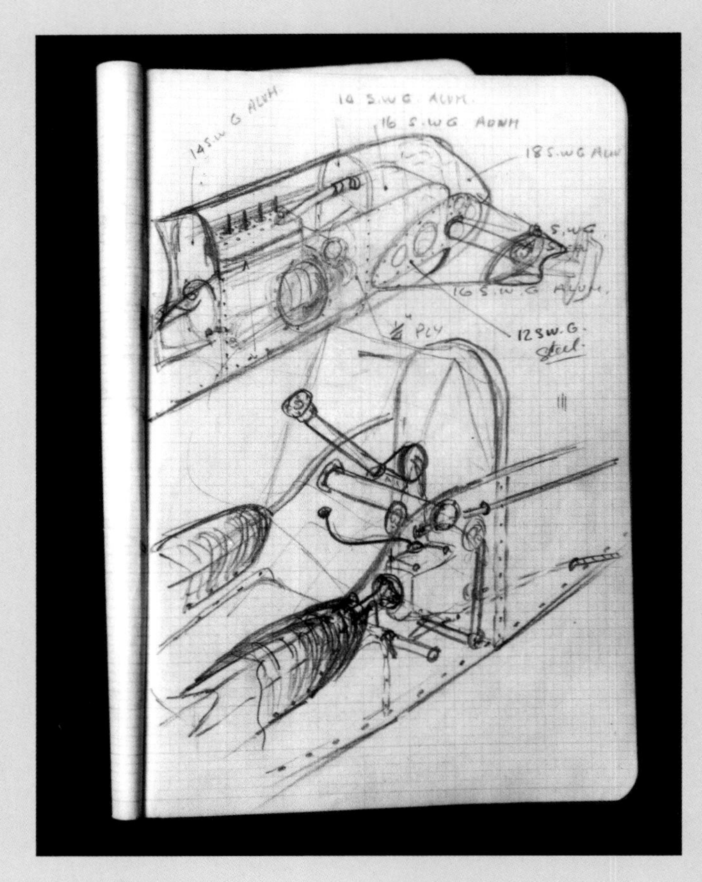

An original page from one of Alec Issigonis's very many notebooks, in this case showing some of the restless creative energy that he poured into his Lightweight Special racing car.

The great man with some of his favorite things, including an early Mini and Morris Minor (left), photographed inside the Longbridge complex that also included the design center nicknamed the Kremlin.

naught after the company axed the expensive venture, and in 1955 he was back at BMC as deputy engineering director.

The short and frenzied development of the Mini saw Issigonis in his element, solving problems while sticking stubbornly to his original concept. Given the urgency involved, his arrogance and dogged insistence on total control were a benefit, although colleagues frequently found him exasperating and often downright rude. His enthusiasm for the challenge, though, was boundless. On paper napkins over lunch or in the little drawing pad he carried, he sketched and calculated constantly, consigning ideas to paper the moment they occurred to him. His team struggled to keep pace with his restless mental energy.

Issigonis moved on immediately to designing Austin Drawing Office No. 16 (ADO 16), the project that became the Austin/Morris 1100; after that came ADO 17, the 1800, together with the related Austin Maxi and Austin 3-Litre.

Issigonis won acclaim in the wider scientific world when he was elected a member of the Royal Society in 1967. However, after BMC became part of British Leyland in 1968, his power base started to crumble, not helped by the thin profits BMC had produced while making his cars. So although he received a knighthood in 1969, his period as director of research and development was brief, and he retired in 1971, taking an enormous Meccano toy construction set with him as a farewell gift from colleagues. After that, the company retained him as a design consultant for fifteen years, but he had very little real influence over its strategy.

Issigonis died on October 2, 1988. He never married, and until her death in 1971 he lived with his mother, Hulda, in Edgbaston, Birmingham. He was an intelligent, witty, and charming individual, fond of a party and any socializing that involved his favorite tipple of gin. While he could be cutting of anyone he considered to be lacking intellectual rigor, Issigonis could also be disarmingly frank about himself. On his eightieth birthday in 1986, he was asked whether he regarded himself as an engineer, scientist, or architect. "An ironmonger," came his sharp retort.

underscored Issigonis's minimalist instrumentation in the form of a single, large circular dial combining speedometer, fuel gauge, and warning lights for oil pressure, battery, and headlamp full-on beams. Rather than being embedded in a dashboard, this sat in the center of a full-width shelf, with space on either side available for holding gloves, bags, parcels, and books. Below this were the only switches the driver needed: two toggles to activate windscreen wipers and lights.

Understanding Issigonis was key to a successful working relationship, and his team was close-knit, including longtime colleagues Jack Daniels, Chris Kingham, and John Sheppard. They worked together in the Kremlin, a nickname for the engineering center at Austin's Longbridge factory near Birmingham, and together and independently they contributed to dozens of small but crucial aspects of the ADO15 design. For example, with space at a premium under the hood, the cooling fan was placed at the side of the engine. There was no space under there for the battery, so it was located in the boot, with a long cable running the length of the car in a dedicated channel to connect it. This arrangement also helped hugely with weight distribution, making sure the car wasn't overly nose-heavy. A starter button was consequently provided on the floor as the most convenient, and cost-effective, position to cut the feed from the battery. Meanwhile, a valve to limit pressure to the back brakes helped prevent the car from locking up during sudden braking on steep slopes.

The second of the XC9003 prototypes to be completed, with the engine in its initial position with the inlet and exhaust at the front and a hood opening down to bumper level.

Apart from the drivetrain itself, the other factor giving ADO15 its great roadholding and handling potential was its torsional strength. The bare bodyshell weighed just 309 pounds, but its exceptional stiffness was provided by two sills extending from front to rear, a lightweight tunnel in the middle of the car housing the load-bearing exhaust system, and the wheel arches. Crosswise, the robust bulkhead between engine compartment and passenger cell, a strong crossmember beneath the front seats, and the rear bulkhead leading to the luggage compartment all contributed further to this stiffness and, hence, permitted thin roof pillars and large windows to make the car light and airy.

This ghosted side elevation gives a crystal-clear view of the clever packaging of ADO15 that Issigonis planned for his 10-foot-long car; it shows the near 80 percent of its footprint given over to passenger and cargo space.

Prototypes were vivid performers; actually, they were a little too fast. The engine, a four-cylinder, three-bearing-crankshaft, overhead-valve unit, was the 948cc A-Series straight out of the Morris Minor 1000. It produced 37 brake horsepower at 5,500 rpm, which gave the lightweight 1,323-pound car too much power for its brakes and suspension to rein in—and a 93-mile-per-hour top speed.

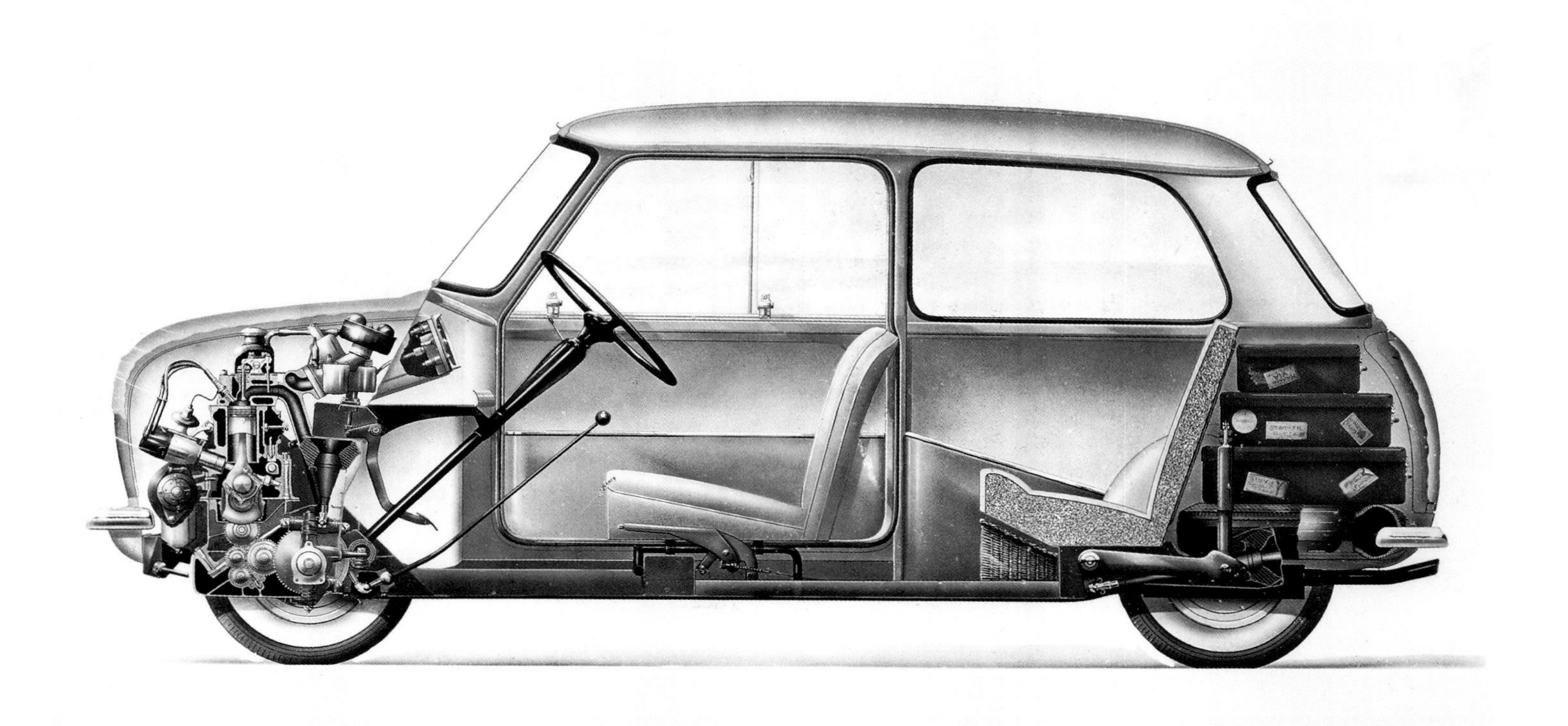

Issigonis invited Lord in July 1957 to come for a spin around the factory grounds in the first prototype. Lord had a try, and then Issigonis took the wheel. "We drove round the Plant, and I was really going like hell," Issigonis recalled later. "I'm certain he was scared, but he was very impressed by the car's roadholding—something that could never be said of other economy cars of the era. So when we stopped outside his office, he got out and simply said: 'All right, build this car.'"

Issigonis and Lord decided to go for a smaller capacity, 848cc, giving a more modest 34 brake horsepower and more leisurely acceleration, with a 72-mile-per-hour top speed, in line with what the small drum brakes behind the 10-inch wheels could be expected to cope with.

As a measure of how fast the design team worked, wooden mock-ups of the car had been completed by July 1957, and a mere three months later prototypes were running, with all Issigonis's hand-drawn sketches now turned into workable engineering drawings. It was an incredible feat to turn all this clever thinking into reality in just seven months. But Issigonis, a walking amalgam of dogma and excitement, made it happen.

Issigonis was adamant that the bodyshell would be manufactured using external welded seams rather than conventional, concealed internal ones, together with external door hinges. The thinking was to free up a tiny bit more space inside and

[above] Alec Issigonis with his three front-wheel-drive cars that redefined motoring in the 1960s: left to right, the Austin 1800, the Austin 1100, and the Austin Mini. Morris editions, of course, were also offered.

[opposite top] A Morris Mini-Minor artfully sliced in half by the British Motor Corporation to demonstrate a cross-section of the car's brilliance; it gives a good idea of the excellent torsional rigidity made possible by the thin pillars.

[opposite bottom] This time the photographer's assistant has filled the Mini's many storage areas with luggage, including sliced-in-half suitcases, a shopping bag in the rear bin, and a small wicker hamper under the rear seat.

save money and time, although it's debatable whether economies were ever achieved because of this, especially after automated body construction became the industry norm. However, these ADO15 attributes became highly distinctive characteristics. Like just about every visual aspect of the car, they were dictated entirely by logic and practicality, for Issigonis was fiercely against styling or anything he regarded as design frivolity.

In fact, the whole exterior look of the car was Issigonis's work, transferred from his sketches and modified to his precise request throughout the building first of the mock-ups and then the drivable cars. The very first such prototype, called the Orange Box after its body color and wearing an upright grille from an Austin A35 on a hood that opened right down to bumper level, was modified during 1958 with a low, wide air intake and a separate opening conventional hood above, sealing the ADO15 look that would become world famous.

There were still some changes to be made, though. The 848cc engine was turned around so that its inlet and exhaust plumbing was at the back instead of the front. But what the team had achieved in just twelve months was remarkable. With the specification coming close to sign-off, eleven proper prototypes were carefully constructed, the last six of which were considered genuinely preproduction material.

Then, in April 1959, two production Minis were hand-built on what would become the new Austin Seven production line in the vast Longbridge plant; one month later, the first ten Morris Mini-Minors were built at Morris's Cowley factory. By June, manufacture was properly underway, with one hundred cars a week being fed into the two BMC dealer networks so there would be plenty of stock for when Britain's most exciting small car ever would be revealed to the world on August 26, 1959.

CHAPTER

MEETING ITS VERY CURIOUS PUBLIC

credible AUSTIN SE

LAUNCHING THE MINI WAS A COMBINATION OF CAR-INDUSTRY TRADITION AND BOLD THINKING. INITIAL RESISTANCE TURNED QUICKLY TO ADULATION. THIS IS HOW IT WAS MADE, HOW IT WAS SOLD, WHAT WAS ON OFFER, AND HOW IT CHANGED THE THINKING AT RIVAL COMPANIES.

Wednesday, August 26, 1959, was the unveiling of what had hitherto been known as ADO15 and was now revealed as the Austin Se7en and Morris Mini-Minor.

The names were picked to draw on as much goodwill as possible from the most famous models from both marques, with a numeric gimmick for the Austin version calculated to catch the eye. "Wizardry on wheels" was the Morris marketing slogan, while Austin used "The incredible Austin Se7en," which was emblazoned on a gigantic pretend top hat that opened to reveal the car, people, and luggage for a cheesy magic-themed launch event at the Longbridge factory, where it had been conceived.

It was the almost incidental use of the prefix *Mini* that stuck, right from the start.

The launch had been moved up from September to steal a march on two other important British small cars due to be unveiled around then: the Ford Anglia 105E and Triumph Herald. Eagle-eyed car spotters might have already seen preproduction Minis out testing between Birmingham and Oxford, or even being sneaked onto dealers' premises. But for most of the British public, the British Motor Corporation's small-car duo was an utter culture shock—as different from the Austin A35 and Morris Minor as it was possible to be while staying on four wheels.

A few days before the public debut, the media had been

Storage in funny places: the Mini's numerous stowage spaces included these useful corners under the back-seat cushion.

invited to meet the cars at the British Army's top-secret military vehicle proving ground at Chobham, Surrey. It didn't take long for journalists to realize that here was something exceptional. John Bolster of *Autosport* magazine, for example, was lavish in his praise of the car's road manners, commenting that Issigonis had "built more safety into this 850cc BMC car than is possessed by any other small machine." *Country Life* magazine noted, "These new models undoubtedly represent a completely new approach to the problems of designing utilitarian economy cars," before telling its readers that Issigonis and his team had beaten even the celebrated Volkswagen and Citroën for sheer design logic.

[above left] A very early Mini stops for breath on a trip around southern Europe, here at the foot of Gibraltar's famous rock.

[above right] A revealing view of the Mini's undercarriage showing the bolt-on subframes and the battery feed and exhaust running down the central strengthening channel.

Readers of *The Autocar*, meanwhile, were treated to the first in a four-part assessment report that far exceeded anything else published in that last week of August. Two journalists had taken an Austin Se7en on an epic 8,197-mile road test from London all around the Mediterranean and back again, hammering the car through European and Middle Eastern countries in a frenzied week.

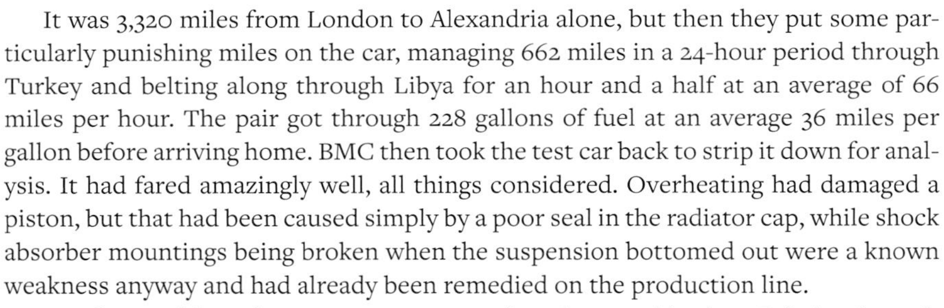

It was 3,320 miles from London to Alexandria alone, but then they put some particularly punishing miles on the car, managing 662 miles in a 24-hour period through Turkey and belting along through Libya for an hour and a half at an average of 66 miles per hour. The pair got through 228 gallons of fuel at an average 36 miles per gallon before arriving home. BMC then took the test car back to strip it down for analysis. It had fared amazingly well, all things considered. Overheating had damaged a piston, but that had been caused simply by a poor seal in the radiator cap, while shock absorber mountings being broken when the suspension bottomed out were a known weakness anyway and had already been remedied on the production line.

Another problem the testers encountered as they sped back to Britain, through a Europe rather wetter than the countries around the Med, was misfiring. The issue was that, since the car's engine had been turned around, its distributor was now exposed to road grime through the radiator grille slats, and the plug leads would soon suffer. Eventually, specially designed plastic boots covering the distributor and coil would be fitted, solving it, but for now a quick fix was planned, with a simple metal splashguard plate to shield them.

But that wasn't the most troubling thing involving the Mini and water. Many early production models filled up with water when driven through heavy rain. At first, perplexed engineers at BMC tried to fix the issue by pumping body cavities with foam, but eventually they found they needed to entirely redesign the floor structure to stop water seeping in through the overlapping join between the main floor pressing and the toeboard. Finally, the source of surface rust and soggy carpets was isolated. In addition, redesign work was deemed necessary on the constant-velocity joints to eliminate mechanical racket, while the gearbox had weak synchromesh that wore out, requiring a rethink with a new baulk-ring design after numerous gearboxes had had to be replaced, expensively, under warranty. A stronger exhaust downpipe was also required so it could properly act as a load-bearing component.

[top left] The graphic artist perhaps reduced the passengers' stature to give a false impression of interior space on this brochure for the original Morris Mini-Minor.

[top right] Austin's marketing didn't include the Mini label but did include a fancy rendition of Se7en to chime with the British love for the original prewar economy car.

[above] Design brilliance gave the Mini a nimbleness and precision that no one had seen before in a device intended primarily to eke out gas.

[top] An Austin Mini Traveller pursues a standard Austin Mini in early 1960s London, where the cars were ideal for darting in and out of traffic.

[above] A Morris Mini-Minor, in this case a De Luxe, with its opening rear quarterlight windows and chrome-plated embellishments.

These were all serious issues that were costly for the company to remedy. They were, though, mostly a result of the Mini's double-quick development time and the sheer amount of new technology crammed into its 10-foot-long package. Problems, really, were inevitable.

The Austin Se7en and Morris Mini-Minor were nearly identical apart from their radiator grilles, hubcaps, badges, and body-paint color choices: the Se7en was available in Tartan red, Speedwell blue, or Farina gray, while the Mini-Minor came in Cherry red, Clipper blue, or Old English white.

You could buy an extremely antiquated Ford 103E Popular for just £419, so the Mini wasn't quite the cheapest proper car on sale. But at £497 including purchase tax, it undercut pretty much everything else (see sidebar on page 43). Potential buyers inspecting the cars had to get used to the peculiar economies that came with this low-price territory. For a start, luggage room within the trunk was tight at a mere 6.83 square feet, but the bottom-hinged trunk lid dropped down to form an open, level luggage platform for bulky items, with a top-hinged number plate that swung down so it remained visible to following traffic. Inside, there were no door handles—Issigonis had provided an adequate, simple, and austere cord to slam the door shut.

Drivers accustomed to the sheer face of a dashboard were surprised at the full-width shelf with a central instrument pod (an astute move intended to save costs

when building left-hand-drive Minis). This was where you put your handbag or hat. There was more stowage space in the deep bins built into each door and beside each rear seat, and they could hold a lot of stuff. Issigonis once playfully suggested the four pockets on the Mini were designed to contain the ingredients, in exact proportion, for his favorite cocktail—a dry martini. To him, that was a 27:1 mixture—twenty-seven bottles of gin and one of dry vermouth.

Dappled sunlight from the roof as an Austin Se7en nears the end of its assembly process; those are Austin A40s taking shape alongside it.

The Mini made a giant impact on the car world, a symbol of Britain's industrial innovation resurgence after notable failures such as the Comet jet airliner. Some have claimed it was almost a failure in its first two years, with just 19,749 cars built in 1959 and 116,677 in 1960, its first full year. It's certainly true that the 1960 figure fell well short of BMC's annual projection of 200,000, but then again one needs to consider the teething problems that could only be solved once the production lines were moving. And it was still substantially more than the 75,000 Austin A35s that were sold yearly.

Still, the Mini could not be judged an overnight success. In addition to problems at the factory, there was some initial resistance to a product so different—the shock of the new. Issigonis's insistence on design purity resulted in a car with no chrome excesses, no aircraft-inspired fins, and a tight, fit-for-purpose attitude to mandatory items such as bumpers and lights. The front windows were of the simple, old-fashioned sliding type, and because Issigonis loathed listening to the radio on the move there was nowhere to install one!

THE COMPETITION

Ford, with its Merseyside-built Anglia just out at £589 and the Cortina development project in full swing, was particularly anxious to understand how BMC had achieved such spectacular consumer value with the Mini's £497 price tag.

[top left] The stylish Triumph Herald, with its Italian lines, was another important debutante in 1959, although it was £200 more costly than the cheapest Mini.

[top right] The rear-engined 1963 Hillman Imp had some thoughtful features, such as this hatchback rear window, but was never very popular.

[left] Fiat's 500 offered minimal motoring Italian style, although its four-seater accommodation was tight and its twin-cylinder rear-mounted engine was noisy.

It bought one of the earliest Minis and took it back to its Dagenham headquarters, where experts in product planning and component buying carefully took apart its 3,016 individual pieces with forensic care. In doing so, they uncovered the Mini's Achilles's heel: its list price was found to be about £30 less than Ford calculated it cost BMC to manufacture the car.

BMC had set the price at this unrealistically low level to win over otherwise skeptical customers. A look at the sums its rivals asked for gives an idea of how keenly the Mini was priced: both the Triumph Herald and the Volkswagen 1200 Beetle were £702, and even the cramped and noisy two-cylinder Fiat Nuova 500 was £499 (albeit including import duties).

In reality, the vast majority of buyers opted for the Se7en or Mini-Minor Deluxe version at £537 with a host of trim upgrades and quite a few genuine improvements, such as a passenger-side sun visor and rear side windows that were hinged for much-needed ventilation. Even then, though, the margins were wafer thin; as the British Motor Corporation chased volume throughout the 1960s, this scarcity of profits would quietly erode the company's prospects, finally forcing it into a merger with British rivals as the industry consolidated.

[right] The start of the Mini's groovy period: this striking op-art Mini decorated with stick-on plastic was for sale in a Thames Ditton showroom in 1966 priced at £385. The model's complementary Shubette dress cost 79 shillings 11 pence and her Saxone slingback shoes 59 shillings 11 pence.

[bottom left] A Morris Mini-Minor Super De Luxe in Oxford in 1963; amusingly, Coopers of Oxford is advertised on both a bus and a van in this picture, although it had nothing to do with performance cars.

[bottom right] A Morris Mini-Minor alongside one of its main British rivals, the Ford Anglia 105E, also launched in 1959 and shown here in estate form.

Dealer networks sold Minis under different brands; this is the cluttered showroom of Austin agent Wades of Worthing circa 1964, with three Austin Minis in stock.

Maybe it was the perception that the Mini was all about bare-bones economy and nothing else that saddled it with a stigma. In the United States, 1959 was notable for the longest, widest production cars of all time, garlanded with some of the most excessive and phony styling flourishes yet seen in the automotive world. The contemporary Ford Anglia 105E, indeed, encapsulated some of that vain glory in its looks and was also a notably responsive car to drive, perhaps giving it and the Italian-styled Triumph Herald more obvious showroom appeal.

The Mini driving experience, however, was like nothing else. Its front-wheel drive, supple ride, and incredible tenacity-to-tarmac handling brought sheer go-kart sensations to the streets. It could be hurled into bends and never lose traction or be pitched off course; its handling limits and roadholding were truly extraordinary, and it almost felt like its door handles scraped the road as it cornered. The surefootedness of front-wheel drive and the racing car–like sensation of having a wheel at each extremity endowed the Mini with astounding nimbleness. The Anglia felt lively, but the Mini was thrilling—swift across country and super-agile in town. One town in particular: London Town.

Once sophisticated city dwellers discovered the Mini in 1960 and 1961, its currency swelled enormously. Nothing got you through the capital's ancient, jumbled streets faster. It could then be squeezed into tiny parking spaces, mews garages, or front gardens. You didn't need to be hard up to appreciate the car's excellent urban characteristics, nor to love its cheeky looks and honest charm. London's smart set adopted the Mini, and then it really took off. At the top of the pile was Lord Snowdon, celebrity photographer and husband of Princess Margaret. He was an avid Mini owner and became good friends with Issigonis. It was through this link that Issigonis took a Mini to Windsor Great Park in 1960 so that the Queen, quite a keen driver, could try the car for herself. The designer must have been giddy with the experience, as he was a bit of a snob himself and henceforth thrived on his society connections. There was nothing so vulgar as a royal endorsement, of course, yet as the tumultuous 1960s got underway the Mini quickly became the "it car."

CHAPTER

COOPER'S MAGIC

AUSTIN COOPER
DPB 760B

IN 1961, THE HOT MINI ARRIVED AND CHANGED THE WAY EVERYONE THOUGHT OF A COMPACT PERFORMANCE CAR ON ROAD, TRACK, AND RALLY COURSE. A GIANT KILLER IN EVERY SENSE OF THE WORD, THE NAME MINI COOPER ENTERED THE LEXICON AND DEVELOPED ITS OWN CULT.

Alec Issigonis always vehemently denied that high performance entered his thoughts as he designed the Mini. But there was no doubt, thanks to its front-wheel-drive concept, low center of gravity, and clever weight distribution, that the car was manifestly far safer than rivals when driven fast. "We were preoccupied in the design with getting good roadholding and stability, but for safety reasons, and to give the driver more pleasure," Issigonis said.

The design maverick had enjoyed considerable sporting success with his Lightweight Special, the supercharged single-seater he built in his spare time with friend George Dowson. Completed in 1938, the car gave a pasting to similarly engined Austin Sevens on the Prescott hillclimb; after the Second World War the two friends evolved it further with an overhead-camshaft engine conversion, and the beautifully finished car continued to impress all who witnessed its giant-slaying powers. Finally, Issigonis's bosses at Morris forced him to give up racing. but that was not before, at the 1946 Brighton Speed Trials, he had met fellow constructor-driver John Cooper, whose 500cc Cooper Special was contesting the Formula 3 class and who proceeded to beat the Lightweight. The two became firm friends.

Zooming forward to 1959 again, and to the lofty amusement of all who witnessed it, the Mini started to appear in a few minor motorsport events, mostly sprints and driving skill tests. With its meek 34 brake horsepower from an 848cc engine, outright acceleration was hardly a forte, but the car's nimbleness and the precise way it handled were immediately obvious. Patronizing smiles dissolved from onlookers' faces.

The Mini took its first class win in a circuit race before the year's end at Snetterton, thanks to George "Doc" Shepherd (he was British Touring Car champion in 1960). It was the early days of understanding the understeer characteristic of front-wheel drive, where the front of the car runs wide of the line the driver takes

One of the very first production cars, officially known simply as an Austin Cooper, pressing on somewhat.

as he turns fast into a corner. At that point, deftly touching the brakes, lifting off the throttle, and/or twitching the steering would unseat the rear tires to bring the tail around—and, with power full on, the car could be made to drift spectacularly through the corner in a way that gracefully outmaneuvered rear-drive cars. It was exhilarating to perform and, of course, made for great entertainment for spectators.

As a race technique, it was brand new, and drivers loved it. There were drawbacks, however; the pressure on the wheel centers in hard cornering was like nothing known before, and the alarming number of wheels detaching and bouncing away into the distance led BMC to design stronger wheels that, of course, made all Minis safer. Other points taken to their limits were oil seals leading to slipping clutches and the engine-stabilizing exhaust pipes, which were prone to breakage.

Once BMC and the racing teams were on top of these problems, the Mini's fortunes took a dramatic turn for the better when Sir John Whitmore won the 1961 British Saloon Car Championship, both the 1-liter class and outright driving one. It was the Mini's first big title win.

John Handley was the first person, privately, to rally an early Mini. He bought one on the car's launch day and entered it in early autumn 1959 in the Worcestershire

Rally, where it acquitted itself well to the usual bemusement of his friends and spectators. BMC's own rally team, based at its Abingdon Competitions Department, was preoccupied with the still very successful Austin-Healey 3000 but felt obliged to take an interest in the new baby. So, in September 1959, competitions manager Marcus Chambers took an early production car to the Viking Rally in Norway, where, in defiance of the atrocious roads, it took fifty-first place. In November's very wintry British RAC Rally, things went downhill, and none of the three team cars finished at all, but in the Portugal Rally in December, Nancy Mitchell came home fifty-fourth and Peter Riley sixty-fourth.

All of this, though, was a prelude to the highest-profile rally of all, the Monte Carlo, in January 1960. Here, four of the six works cars that were entered crossed the line, the highest-placed being Peter Riley again in twenty-third place overall. (Another six private Minis also competed.) At the Geneva Rally four months later, works driver Don Morley and his co-driver and brother Erle upped the ante, grabbing the Mini's first class victory and a respectable thirty-fourth place overall.

While all this was happening, John Cooper was riding high. Ever since he'd decided to put the engine behind the driver in his 500cc motorbike-powered racing cars, they had seen amazing success, jumping up from Formula 3 to the top-level

[top] That little boy earning his pocket money is future F1 racing car designer Adrian Newey, and the car is the family Morris Mini Cooper he later learned to drive in.

[above] John Cooper (right) explaining the benefits of a rear-engine configuration in F1 to his young son, Michael; let's hope the briefing was followed by an ice cream.

The cover of Castrol's annual Achievements booklet in 1962 was already casting the Cooper as a hotshot on the rally scene.

Formula 1. In 1959 and 1960, Coopers won the F1 World Championship, and his cars had by then helped launch the careers of Stirling Moss, Graham Hill, Jack Brabham and Bruce McLaren. Cooper was a great advocate of using near-standard production engines in his Formula Junior single-seaters, and BMC's A-Series was one of his favorites. Both McLaren and Brabham had bought Minis, and Cooper could see the obvious potential for more power. He suggested the idea to his old pal Issigonis, and it quickly received the enthusiastic blessing of BMC managing director George Harriman. A mass-produced "hot" version of no more than one thousand units, argued Cooper, would qualify a more powerful Mini as a catalogued model and therefore allow it to take part in motorsport as a production car.

Cooper increased the capacity of a standard 848cc engine block to 997cc, reducing the bore from 62.9 to 62.4 millimeters and enlarging the stroke from 68.3 to 81.3 millimeters. The compression ratio was raised from 8.3:1 to 9:1. Larger intake valves

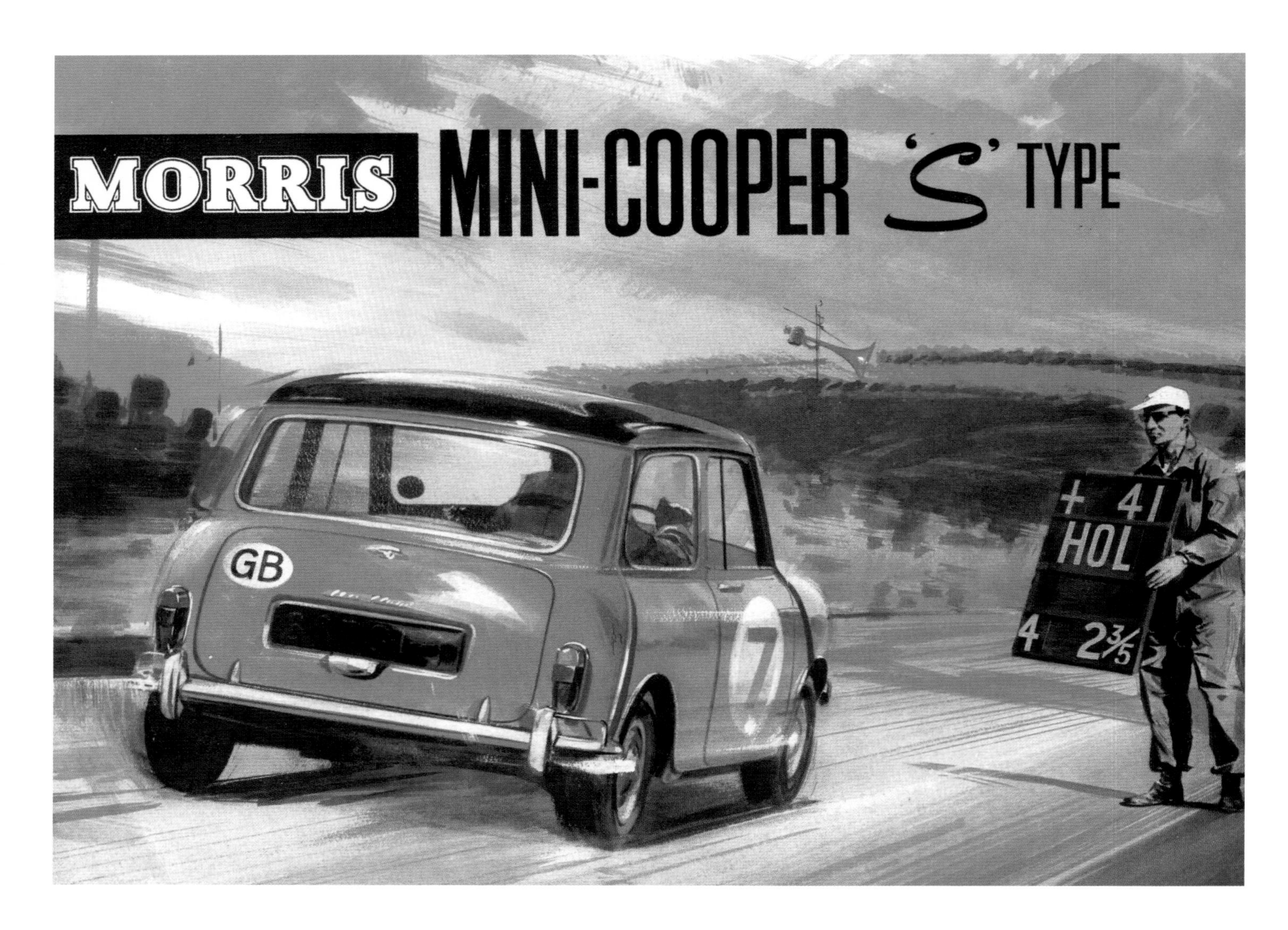

[above] Lovely artwork on the cover of the first Cooper S brochure left potential buyers in no doubt that this was a car begging to be raced on the weekend.

[right] The Morris Mini-Cooper S Mark II brand new in 1966, when it was almost certainly the world's most desirable small car.

Not every Cooper that tackled the Monte covered itself in glory; this is the Claude Twigdon/Anthony Gorst Cooper S going great guns before a forced retirement from the 1965 rally.

and exhaust bore on the three-branch manifold were specified, and twin SU carburetors were fitted. More fuel and air getting in faster meant more power coming out, just as in a Formula Junior engine. The crankcase was reinforced so the engine could withstand the substantial power hike.

To improve the gear change—with closer gearbox ratios for better acceleration in each gear—Cooper swapped the long, wavy gear lever for a snappy remote item. But the key change was to work with Dunlop to produce front disc brakes, at 7 inches in diameter the smallest in the world, to handle the significant leap from 34 to 55 brake horsepower.

The transformation was sensational, with acceleration from 0 to 60 miles per hour slashed from 30 to 18 seconds and maximum speed jumping from 73 to 88 miles per hour. John Cooper dispatched F1 driver Roy Salvadori to the Monza test circuit in Italy in a prototype for some sustained high-speed track testing in 1959. Salvadori reached the destination an hour before his colleague Reg Parnell . . . who was driving an Aston Martin DB4.

CONTINUED ON PAGE 57

MONTE CARLO AND THE COOPER S

With the Cooper's upgrade to S level, BMC now possessed a car that could take outright victory, not just class honors. And so, in January 1964, Paddy Hopkirk and co-driver Henry Liddon drove their Cooper S, registration number 33 EJB, to finally win the Monte Carlo Rally for the first time in giant-killing form. Starting from Minsk, in Russia, the red-and-white car put powerful Citroëns, Mercedes-Benzes, and Ford Falcons in the shade. Timo Mäkinen's Cooper S was fourth, Rauno Aaltonen's seventh, and unsurprisingly, BMC took the Manufacturers' Team Prize.

Mäkinen and co-driver Paul Easter would swipe another Monte Carlo win in 1965 in one of the new 76-brake-horsepower, 1,275cc Cooper S cars, registered AJB 44B, an epic drive underscored by the fact the duo incurred not one penalty point despite some of the snowiest conditions ever seen on the thousands of miles of the event. Only 35 of the 237 cars entered made the finish, and 3 of those were Mini Cooper Ss.

Mäkinen was set to make it a Cooper S hat trick in 1966, but his winning car, GRX 555D, was promptly disqualified (along with Aaltonen's second-placed car and Hopkirk's in third) for a dipped-beam headlamp system involving coordinated headlights and spotlights that contravened regulations. A Ford Lotus Cortina was also denied its fourth place. The questionable victory was handed to a Citroën, and the British press managed to whip the controversy into a matter of national importance that got the whole country indignant at the apparent injustice . . . and French arrogance.

But the three musketeers were back, undaunted, in 1967 and dominated the show in style. LBL 6D, driven by Aaltonen and Liddon, won the rally to tumultuous applause from spectators; Hopkirk came home sixth and Mäkinen forty-first.

In 1968, the rankings started to slip, with Aaltonen third, Tony Fall fourth, and Hopkirk fifth in the Monte. But, then again, after 263 event entries that had produced 109 different trophies, there was little more to prove, as a new generation of powerful Porsche and Alpine-Renault sports cars came to prominence and highlighted the Mini's weakness of small wheels whose tires and brakes heated up and wore away more quickly than those on rivals' cars. On a personal level, Aaltonen was the most successful Mini Cooper rally driver; he won the 1965 European title and tasted eight outright wins and fourteen class wins. Stuart Turner, BMC competitions manager from 1961 to 1967, later recalled that it was the "right car at the right time"; the Mini's showing confirmed it as among the most significant rally cars of all time.

[top] The 1964 Monte started in Minsk, and Hopkirk and Liddon beat off all their more powerful competitors to take a brilliant victory.

[above] We was robbed: officials pointing out the faulty headlight setting that denied a 1966 Monte Carlo win to Timo Mäkinen's Cooper S.

[right] UK heroes Hopkirk and Liddon receiving the '64 winner's trophy in Monte Carlo.

[opposite] Hopkirk and Liddon scattering gravel on the 1964 Monte Carlo Rally in 1964.

[top] The team of Fall/Wood on the '68 Monte, in which they came home fourth.

[above] The Cooper S of Hopkirk/Crellin that came home an excellent fifth on the 1968 Monte, with its iconic roof rack and spare wheels.

[left] Timo Mäkinen and Paul Easter drove faultlessly to win the epic 1965 Monte, one of the wintriest ever known. Only 35 of the 237 cars that entered actually made the finish.

This Austin Mini Cooper S. driven by Simo Lampinen and Tony Ambrose was forced to retire from the 1966 RAC Rally in Britain.

Despite some misgivings that BMC might struggle to sell them all, Harriman gave the green light to make a limited series of one thousand cars, with a £2 royalty on each one in exchange for being allowed to call it the Mini Cooper.

It was launched in September 1961 in all-but-identical Austin and Morris liveries at £679 and was immediately recognizable for its two-tone paint job, with a contrasting roof, and wheels drilled with holes around the perimeter for extra brake ventilation. Demand was huge almost from the start, and all thoughts that this would be a limited-run model were soon forgotten. At its peak, BMC was making 750 of them a week. And the Cooper factor—with a nearly 50 percent power boost for a minimal weight penalty—had a truly transformative effect on the car's competitive life.

For one thing, it started to actually win rallies, beginning with Pat Moss's victory on the Tulip Rally in the Netherlands in May 1962 (see sidebar on page 59) and culminating in third (and class victory) and sixth overall for Rauno Aaltonen and Paddy Hopkirk.

Then again, the Mini Cooper had a blinding first full year on the racetrack.

Class victories at home were clinched at Aintree, Brands Hatch, Crystal Palace, Goodwood, Mallory Park, Oulton Park, and Silverstone, with Mini Cooper driver John Love winning a British Saloon Car Championship that made fools out of

Cooper S sets the track alight at Oulton Park in 1965; drivers such as John Rhodes became the new heroes of racing as they juggled speed and nimbleness.

Jaguars and Ford Galaxies. Meanwhile, further afield, the cars won everything from the saloon-car race supporting the German Grand Prix to a class win in the 12-Hour Sedan Race in Washington D.C., USA, and a 1-2-3 in the Swedish saloon racing series.

It was clear to John Cooper that there was even more performance potential to be extracted from the Mini. But it took motoring journalist Ronald "Steady" Barker to convince Issigonis. In an interview with renowned historian Jon Pressnell, Barker recalled the telephone call he made to the design chief in 1961. "I said, 'I've just done something over 100 miles per hour in a Mini.' He couldn't believe it. He said, 'Is there any chance of bringing it up here?' So I went up to Longbridge and he was just like a schoolboy with a new toy. He kept giving it bursts of acceleration, smiling like a demon."

The car featured a bespoke 1088cc engine tuned by Daniel Richmond of Downton Engineering and was remarkable for its smoothness and excellent low-speed torque, which made it docile for everyday use yet fiercely responsive when pressed. It was so accomplished that Richmond was hired to oversee development of the car that would become the Cooper S, launched in March 1963.

At its heart was an engine with capacity increased once again to 1071cc to keep it within 1100cc race class regulations by widening the bore as far as possible within the engine block but keeping the stroke as short as on a standard 848cc engine.

PAT MOSS

Within weeks of the car's launch, Pat Moss—sister of Stirling, later wife of Swedish rally legend Eric Carlsson—showed that the basic Mini could be a rally winner, albeit in a minor British club event in Lancashire. At the end, Pat described her winning Mini as too slow—and co-driver Stuart Turner thought it far from comfortable.

Legendary rally driver Pat Moss getting her hands dirty with one of her famous Big Healeys; she was first to give the Mini Cooper a rally victory on the 1962 Tulip Rally.

Moss became the most successful of the many female Mini works rally drivers. Indeed, she made headlines in May 1962 when she gave the 997cc Mini Cooper its first ever outright rally win on the Tulip Rally in Holland.

As Stirling Moss's sister, keener on horses than cars, she might have begun as a publicity-gaining novelty, but Pat soon proved to be a formidable member of the BMC stable. She took a Morris Minor to fourth place on Britain's RAC Rally in 1958 before moving to Austin-Healey 3000s, in which she won outright the 1960 Liège-Rome-Liège and took a second on the Coupe des Alpes.

Later, she drove all manner of cars, from Lancias to Porsches and Fords to Saabs, but few of her always creditable results was more impressive than her breakthrough in the Mini Cooper—the car she regarded with suspicion as "twitchy, and pretty unruly."

Think of a performance-car cockpit today and then look again at the positively barren driving environment of this Mark II Cooper S 1275.

Stronger pistons, valves, and valve gear were used too, while a nitrided crankshaft featured bigger main bearings. Output was 70 brake horsepower, more than twice that of the original Mini.

A strengthened clutch and gearbox were part of the package too, along with thicker, slightly larger 7.5-inch disc brakes with vacuum-servo assistance and wider wheels with radial tires. The brakes were one of the key features of the car because the S took a quantum leap in performance. Top speed was a remarkable 95 miles per hour, and it could positively scream from standstill to 60 miles per hour in 13 seconds. The price was £695.

Shortly afterwards, in January 1964, the profitable standard Mini Cooper came in for a rethink. It gained the basic engine from the Mini-based Riley Elf/Wolseley Hornet cars, which was much stronger and more receptive to tuning. Power output was the same, at 55 brake horsepower, but there was more torque.

Also in 1964, the Cooper S engine choice expanded to three, with 970 and 1,275cc versions joining the 1,071cc model. This was part of John Cooper's scheme for the trio to allow Mini Coopers to contest three saloon car race classes: up to 1,000, 1,100, and 1,300cc. They shared an identical bore with different strokes, from the short 970cc to the long 1,275. This last engine was the most flexible and least fragile; with its 76 brake horsepower, it also finally edged the Mini into 100-mile-per-hour territory, with a fine 79 pounds-feet of torque to make it brilliantly versatile in the midrange. Acceleration to 60 miles per hour was 11.2 seconds. It survived right up to the end of the original Mini Cooper era in 1971. The overlap period was very brief: the 1,071cc model was dropped in August 1964, while the mercurial 970 lasted only from June 1964 to April the following year—after just about enough cars had been built to meet the racing homologation requirements. It was the 1,275 S that everyone wanted and loved.

The Mini Cooper S became an automotive highpoint of the 1960s through rallying (see sidebar on page 59). This was what the factory concentrated on, while supporting private teams who took the cars circuit racing. Cooper's own équipe adopted the same evocative livery as its Formula 1 cars, British racing green with white stripes. John Love and John Fitzpatrick won their respective 1962 and 1964 British Saloon Car Championships titles under the Cooper aegis. Dutchman Rob Slotemaker scooped the European 1,300cc title in a works Downton car in 1963, and the year after Warwick Banks won both the British and the European 1-liter classes.

Close-fought saloon-car racing action at Silverstone in 1965, where three battling Coopers provide much more entertainment than the average support race—and all without roll cages.

In 1965, it was John Rhodes who won the British 1,300 class driving for Ken Tyrrell's team. His style was furious, wheel-to-wheel action with his Mini-driving rivals like Ralph Broad and Gordon Spice, and smoke pouring off his tires as he jostled to get ahead. His superbly controlled drifting skills were the best, and there was nothing so spectacular to be seen anywhere in the field of "normal" cars racing one another. Rhodes's Mini Cooper mastery saw him achieve the unprecedented feat of winning the British Saloon Car Championship 1,300cc class all four times between 1965 and 1968.

The Cooper S's racing swan song came in 1969. The works this time campaigned its own single team of cars, driven by Handley and Rhodes, and supported no others, but it didn't get close to taking overall championship honors. While the sun was setting on the Mini's racing career, however, there was still Richard Longman to give it a fillip by winning the RAC British Saloon Car Championship in 1978 and 1979 (in a 1275GT, rather than a Cooper). And as the Mini's glory days were gently fading, the drivers whose careers it helped launch—Jochen Rindt, James Hunt and Niki Lauda—were heading for Formula 1 stardom. Lauda started racing a secondhand Cooper S while vainly trying to keep his antics a secret from disapproving, fearful parents; he came third in it in the very first race he entered in April 1968. Hunt, meanwhile, worked as a hospital night porter to fund the building of his first racing Mini in 1967, which was soon sold to buy a Formula Ford single-seater. These days, F1 prodigies start out in karts, but back then, it was usually Minis.

CHAPTER

A WORLD OF DIFFERENT MINIS

AUSTIN
CRANE FRUEHAUF
BMC

CHOICES FROM THE FACTORY EXPLODED, BUT THAT WAS BEFORE EVERYONE ELSE DECIDED TO INTERPRET THE MINI THEIR OWN WAY. VANS, PICKUPS, AND THE MOKE MODEL PUT MINIS INTO EVERY SPHERE OF WORKING LIFE, AND THE BASIC CAR WAS MANUFACTURED ALL OVER THE WORLD.

[above] The Mini van captured in artwork for a promotional postcard sent out by dealers to every local small business they could think of, confectioners included.

[opposite] The RAC was one of many big fleet customers for the new Mini van, using them to replace old motorcycle-sidecar combinations hitherto used by its breakdown patrolmen.

Well before the Cooper started to shake up the world of performance cars, the Mini lineup had started to burgeon. The new derivatives came thick and fast as BMC, in the typically thorough manner of a company that manufactured just about all types of motor vehicle apart from motorbikes and tanks, sought to wring every last bit of value out of the basic design.

The first group to be targeted was business customers. In May 1960, wraps came off the Austin and Morris Mini vans. The 9.5-inch increase in overall length was immediately obvious, but the steel-paneled sides also concealed a 4-inch-longer wheelbase. The box-hungry 46 cubic feet of cargo capacity was accessed through a pair of side-opening back doors. Clever design put the spare wheel and battery under the floor within the wheelbase and the fuel tank below it behind the rear wheels. The vans were immediately popular, with fleets several hundred strong ordered by the post office and the Automobile Association, where they started to replace the motorcycle/sidecar combinations previously forced on patrolmen.

Four months later, the Austin Se7en Countryman and Morris Mini-Minor Traveller made their debuts—two pint-size estate cars using the van structure and rear doors but with sliding side windows and a rear seat that folded forward to leave a still cavernous 35 cubic feet to fill with chattel. For a family resemblance to the Morris Minor Traveller, the rear body section featured a half-timbered look, although unlike on the Minor the varnished wooden trim was decorative rather than the

actual structure supporting the rear body panels. The overall look was very charming, and the Countryman/Traveller duo was particularly popular in France, where they made chic Parisian town cars.

Rounding out the long-wheelbase lineup was the elfin Mini pickup of spring 1961. With a drop-down tailgate, one of these could carry a quarter of a ton of cargo, just like the vans.

Early on, BMC's thoughts had turned to offering a Mini with a luxury touch, and autumn 1961 witnessed two different approaches. The first was the Super, which was little more than the Cooper without its twin-carb engine and disc brakes but with the sporty car's two-tone paintwork, three-dial instrument panel to include oil and water gauges, overriders with tubular corner bars, and twin-color interior trim. It was only around for a year before being merged with the De Luxe to form one Super De Luxe option.

This latter move was probably a result of the other late-1961 debutantes, the Mini-based Riley Elf and Wolseley Hornet duo, which really did represent a leap upmarket. Resplendent in the liveries of these two stablemate BMC marques, including upright traditional front grilles and smartly trimmed interiors with part-leather covered seats, they were more practical than the standard Minis because of the boxy extended trunk at the back with its much-enlarged capacity. Bracketing the trunk lid were tiny rear fins that echoed the style of BMC's larger cars, while the Riley also included a full-width wood-veneer dashboard. Both offered an attractive package, although, curiously, they were for years mocked by diehard Mini fanatics as being elaborate dress-ups of the standard Mini.

[top] Mini vans were soon found in every walk of life, such as on this somewhat rundown-looking farm.

[above] A freshly minted Morris Mini pickup ready for its hard life ahead keeping the wheels of British business turning.

This period 1962 brochure pitched the Traveller at the small family market and took the usual graphic liberty of exaggerating the ratio of people's stature relative to the car's tiny dimensions.

Perhaps one of the reasons for the sneering was that the outwardly opulent duo had to make do with the standard Mini's 34-brake-horsepower, 848cc engine; as they weighed almost 10 percent more than the standard car, performance was severely blunted, with the 0-60-mile-per-hour "dash" taking a lethargic 32.3 seconds. An engine upgrade in 1963 with the Mark II versions of the cars couldn't come a moment too soon, and the added 4 brake horsepower from the new 998cc power unit provided the missing liveliness.

The quest for true Mini exclusivity was almost certainly triggered by obsessive car nut Peter Sellers (see sidebar on page 68) and his money-no-object one-off, but London coachbuilding company Harold Radford capitalized on the idea of creating miniature limousines with the launch in April 1963, of its Mini De Ville in full-spec Grande Luxe form at £1,080, the slightly cheaper Bel Air, or the merely quite fancy De Luxe. Although every car was different according to its individual customer specifications, modifications typical of the bespoke, hand-finished conversions included interiors lavishly trimmed in Connolly leather, Bentley-standard

CONTINUED ON PAGE 70

PETER SELLERS AND HIS MINIS

In a mostly pre-TV era, Peter Sellers's daft personae of Bluebottle and Major Dennis Bloodnok in *The Goon Show* enlivened radio nights for grim postwar Britain. Then the blundering slapstick of Sellers's Inspector Clouseau sustained the phenomenally successful Pink Panther film franchise in the 1960s and 1970s. Sellers made a string of other comedy cinema greats that are still enjoyable: *I'm All Right Jack*, *The Battle of the Sexes*, and *Dr. Strangelove*.

His uncanny skill for mimicry propelled his rise from nobody to 1960s showbiz A-lister. The big money rolled in, and Sellers splurged it on cars, gorging himself on precious metal like few celebrities before or since.

His appetite for the motoring high life was whetted in April 1959 when his brand-new Bentley S1 Continental arrived. Yet that was barely run in when he bought an ex-Cary Grant Rolls-Royce Silver Cloud I followed by a Cloud III fixed-head coupe, and by 1963 he'd branched out into exotica proper, owning an Aston Martin DB4 GT Vantage and a Ferrari 250 GTE. Both featured in his crook caper movie *The Wrong Arm of the Law* that year.

Not surprisingly, perhaps, he started to find his Morris Mini Cooper a little too austere, which gave him the novel idea to have the car professionally and comprehensively customized by respected London coachbuilder Hooper Motor Services, whose many decades in business had seen the firm create countless Rolls-Royce limousines for the House of Windsor and many other royal families around the world. "Anything you boys can think of you have my full permission to do," he apparently said in late 1962.

They took him at his word. Inside, a full leather re-trim was complemented by a full-width mahogany dash with matching door cappings, padded and reclining Reutter front seats, deep Wilton carpets, wool cloth headlining, transistor radio with twin speakers, wood-rimmed three-spoke steering wheel, electric windows, and a more powerful heater. There was a full-length sunroof and twin Lucas spotlights mounted ahead of the grille, but the most distinctive aspect was the lustrous purple paintwork with, on the sides, hand-painted wicker-effect panels, executed by Hooper's in-house heraldic artist, Geoff Francis. The cost was an eye-watering £2,600, four times the cost of a Cooper, and Sellers collected what was then undoubtedly the ultimate Mini in May 1963.

He loved it so much he had a doppelganger created by Harold Radford to appear in his 1964 Clouseau film *A Shot in the Dark*. Sellers gave this one to director Blake Edwards, and then ordered a Mini De Ville GT from Radford, with hatchback, for new bride Britt Ekland; the actress drove it out of a large pretend wedding cake in Radford's Hammersmith showroom in October 1965.

Just days later, the couple visited the Earl's Court Motor Show, where Sellers bought the brown Ferrari 500 Superfast making its British debut there. "I'm keeping it forever," he later swooned. "I don't want anybody else to drive it but me." But this infatuation didn't stop him from ordering a Lincoln Continental, and then a Rolls Silver Shadow two-door coupe in March 1966. Fellow *Goon* Spike Milligan called Sellers's car obsession "metal underwear" because he changed them so often. The AA's *Drive* reported in 1967 that he'd owned eighty-five cars and revealed some of his manic compulsiveness for trade-ins: "Sellers is the complete automaniac. He owned a Jaguar for one day, a Volvo for two, a Rover for three."

Sellers died of a heart attack aged fifty-four in 1980 following a tortured family life—after four marriages, depression, addiction, and a career dip. Shortly before he passed away, though, he wrote, "For thousands of us who had to get around London quickly, the arrival of the Mini was like the answer to a prayer."

Peter Sellers's Radford Mini De Ville, which had a starring role in his 1964 movie *A Shot in the Dark*.

gleaming two-tone paintwork, full-length sunroofs, up-to-the-minute hi-fi systems, and electric gadgets only found on the most expensive luxury GTs, including power-operated windows and radio aerials.

London rivals Wood & Pickett soon joined the fray, and the two companies competed to offer their customers ever more ambitious features. Remodeled front seats became more and more luxurious, while bodywork could be modified to include a hatchback rear door and folding seats. Stylish touches included ridding the body of its exterior seams and rain gutters for a smoother, sleeker overall look, blacked-out windows, and a front end reshaped around Mercedes-Benz headlight stacks. Downton-tuned engines and beefier batteries were often called for to give power to all the expensive handmade opulence.

Among the most famous individual cars was a 1275 Cooper S created in July 1967 by Harold Radford for Mike Nesmith of pop group the Monkees. Finished in metallic sable with champagne-pink upholstery, the interior featured unique seats and instruments together with a built-in tape player/recorder and a one-off ventilation system so it could cope with the heat of California, where the car was ultimately bound. At £3,640 it was said to be the most expensive Mini ever built.

And then there was film star Laurence Harvey's Wood & Pickett Cooper S, completed in 1970 and almost as costly at £3,500. As well as the widened wheel arches, tinted windows, lambswool carpets, full-length Webasto sunroof, Sony cassette player, and VHF radio, Harvey's initials glittered in gold leaf on the doors as part of a lustrous paint job. The Downton-tuned engine was obvious from the rasping bass of the special exhaust system.

Cars like these inspired dozens of small specialist firms to offer a galaxy of customizing accessories to less monied Mini owners. Many offered components that improved the everyday Mini driving experience, such as a remote linkage for a crisper gearchange or a bigger-capacity fuel tank to improve the touring range of the

CONTINUED ON PAGE 74

[top left] The Wolseley Hornet offered sizeable extra trunk space and a prestigious radiator grille, but dedicated Mini fans were always sniffy about the car.

[top right] Left-hand drive for the Mini Traveller—the little wood-trimmed estate cars were much adored by chic Parisians.

[above] The dad owning this early Austin Se7en Countryman looks like he'll soon be missing his hood badge, if junior has his way.

[above] The Morris Mini Countryman was a smart-looking Mini that was also extremely versatile and very thrifty.

[left] A full-width walnut dashboard inside the Riley Elf offered gloveboxes on both sides, although the optional radio looks like something of an afterthought.

THE MINI MOKE

In its ceaseless quest to create new Mini derivatives, in 1959 BMC started to examine the prospects for a light military version. The impetus was the need, identified by Britain's Ministry of Defence, for troop transport that was cheaper and lighter than a Land Rover.

It might sound fanciful now, but the basic concept was for a minimalist vehicle that could be stacked three high inside a military transport plane and then be parachuted into combat situations along with four soldiers for each one. Although it obviously lacked four-wheel drive, the thinking went that the Moke was light enough to be carried over anything too boggy by those beefcake troops.

The Moke name was adopted as early as 1959, and British army chiefs had a fleet of prototypes on trial that year. The reaction was lukewarm. Its low ground clearance and tiny 10-inch wheels were seen as hindrances, although the vehicle impressed with its ability to hop over many small obstacles. The buckboard body shell was made up of the Mini floorpan with wide, box-shaped side sills, together with a simple steel trunk to house the engine and carry a windscreen. The wing-tops were flat to support the wheels of Mokes resting on top of them.

The Royal Navy bought a handful, but the army soon abandoned its real-time assessments. Still, BMC didn't want to waste the car. So it went on sale in January 1964 as an Austin or Morris Mini Moke, with an open-sided canvas tilt/hood and storage lockers built into its sides. Green paint was the only option. The price was just £405, with only a Fiat 500 or a Goggomobil cheaper, by £5, but you had to pay extra for all three passenger seats and even a laminated windscreen and a second windshield wiper. HM Customs and Excise insisted the 65-mile-per-hour Moke was a passenger car and attracted purchase tax, even when sold with just a driver's seat because it had been designed around a four-seater layout.

Consequently, some 90 percent of Mokes were exported and sold as beach cars and hotel taxis in hot countries—it was Britain's eccentric riposte to America's dune buggy. Most of the remainder stayed in 1960s London, where afghan-coated hippies shivered at the wheel as they waited at Notting Hill traffic lights. *Motoring Which?* summed up

The presenter team from the BBC's children's TV show *Blue Peter* in a Moke outside TV Centre in Shepherd's Bush, London, in 1969.

[left] Issigonis, Cooper, and BMC all toyed with twin-engined Minis; this is the four-wheel-drive Twini Moke having a fine time scattering snow and hauling hay bales simultaneously.

[below] The Moke had a surprisingly long life, with manufacture transferring first to Australia in 1968 and then to Portugal in 1983, where this late-period example was built and found a ready market as a holiday resort runabout.

the experience by pointing out: "Driving through the back streets of Kensington in pouring rain in the Moke must rate, as an activity, very low on anyone's fun index." Mokes featured prominently in cult ATV series *The Prisoner*, and in the amazing underground scenes in the Bond movie *The Man with the Golden Gun*. In October 1968, British manufacture ended after 14,518 had been built (all but 1,467 exported), and the entire Moke tooling was shipped to Sydney, Australia. It was manufactured there, and later in Portugal, until 1994.

Issigonis himself built an experimental four-wheel drive "Twini Moke" that was shown to reporters at the time of the Moke's launch and demonstrated by the exuberant design chief. With a 948cc engine at the front and an 848cc unit at the back—and two gear levers—by all accounts it gave an awe-inspiring performance, and in fact in the previous heavy winter Issigonis had had a great time scattering snow with it. Once again, army chiefs spurned the project for its obvious complexity, but the Twini did prompt Issigonis and his pal John Cooper to each build a twin-engined Mini saloon. Although they were purely experimental, Cooper's led on to a twin-engined car that Sir John Whitmore used to contest the Targa Florio road race—extremely quick it proved, with its 175 brake horsepower from two Downton-tuned motors, but impossible to keep cool in the searing Italian heat. However, after Cooper was badly injured in his in a crash on the Kingston bypass in Surrey, the project came to an immediate end.

car. But if anything, there were even more sweeties to choose from if you wanted to boost the performance of your car, Cooper or otherwise.

The back pages of magazines like *Cars and Car Conversions* were crammed with small ads peddling aftermarket performance items. There had been bolt-on performance parts in the past, especially for Austin Sevens and side-valve Fords such as the Anglia and Popular, but nothing like the boom seen for the Mini in the 1960s. It was fueled by the cars' spectacular antics on the racetrack and by the tuning knowledge passed on in big-selling books such as David Vizard's *How to Modify Your Mini* and *Tuning the A-Series Engine.*

Power could be amply boosted simply by changing the exhaust or the carburetors, with even more in store by then improving the cylinder head, lightening the valve gear, and changing the camshaft, pistons, and timing. Uprating brakes, wheels (especially the desirable Minilite and Cosmic alloys), and tires was then almost mandatory so that your hot-rodded Mini could handle its new urge.

Numerous tuning firms flourished in the 1960s, such as Broadspseed, Janspeed, Speedwell, and Taurus. They offered kits or would do the job for you, and right at the peak of this unruly pile was Downton, creator of the Cooper S engine, whose built-to-order engine conversions were among the very best because they were strong and refined as well as giving plentiful extra performance to satisfy even the most demanding of Mini Cooper owners, including Steve McQueen (Downton-engined examples really were 100-mile-per-hour road cars).

Naturally, as the home of sports-car legends MG and Austin-Healey, BMC considered a pure sports car based on the Mini. The idea was proposed several times, with 1960s prototypes such as the ADO 34 roadster and ADO 35 coupe, and the 1970s' ADO 70 as a potential midengined replacement for both the MG Midget and Triumph Spitfire.

[top left] Another hatchback conversion, this time by London coachbuilder Harold Radford, on what is obviously a very fully loaded Cooper S.

[top right] This one-off hatchback Mini Cooper S was built by BMC especially for British minister of transport Ernest Marples as a way to carry his golf clubs and bring wine back from France. Shame it didn't go into production.

[above] The Beatles all had special Minis at one time or another in the 1960s, and this little gem was Ringo Starr's, modified by Radford with a hatchback so its owner could stuff his drum kit in the back.

[above] As the Mini reached its one-million milestone, Alec Issigonis and George Harriman, in the center of the photo, were on hand at Longbridge to congratulate it.

[left] By the time the Mark III version of the Riley Elf was made available in 1966, it had a more powerful engine and became the first Mini with wind-down windows and concealed door hinges.

mini-marcos gt

[above] The Mini-Marcos might have looked a bit odd due to the efforts to make it cheat the wind, but its performance was exceptional and it remained on sale for decades.

[right] There were hundreds of go-faster goodies you could bolt on to your Mini, such as this carburetor upgrade from Stromberg.

None of these came close to going on sale. But that probably didn't matter because for anyone who wanted a Mini-engined sports car, there was plenty of choice. First on offer was the SX1000 from Ogle Design in 1962, a transformation of an existing Cooper into a baby GT with a notably svelte glass-fiber body fitted around the car's original floorpan and scuttle. Some sixty-six were made even though the total price of £1,070 was more than an Austin-Healey 3000.

Then there was the Broadspeed GT of 1966 with its fastback roof, but the Unipower GT of that year went considerably further, with the Mini's front subframe transferred to the rear of the car to produce a midengined two-seater GT rather like a miniature Ford GT40. The radiator remained at the front, but the gear lever was to the right of the driver with a reverse pattern gate and so took some getting used to. The Cox GTM of 1967 was another midengined two-seater that enjoyed an extraordinarily long period on sale.

If you were prepared to build your own Mini-based sports car, then there were many Mini-based kit cars over the years. The best known began life as the DART in 1964—at a time when crashed Minis as donor cars were beginning to be commonplace in scrapyards—but was soon renamed the Mini-Jem. Marcos Cars then introduced its very similar Mini-Marcos in 1966, and this has outlasted the GTM in continuing to be on sale to this very day. The Midas was a very smart update of the Mini-Marcos for the 1980s and was so professionally done that its builder, Harold Dermott, was employed to help make the 221-mile-per-hour McLaren F1 a production car reality in 1992; such are the long tentacles that the Mini has helped extend out into Britain's diverse motoring heritage.

[left] The success of *Cars and Car Conversions* magazine was based to a large degree on Mini fanaticism; it carried reports of Mini racing along with hundreds of adverts for tuning parts and accessories.

[below right] One of the first small GT cars to use the Mini as a base was this nifty Ogle SX1000, launched in 1962, with a fiberglass body.

[bottom] The handsome but short-lived Unipower GT utilized a Mini engine and subframe behind the driver, resulting in quite a weird gearchange.

CHAPTER

EVOLUTION THROUGH THE MARKS

AUSTIN mini
BRITISH PAVILION
1967 MONTREAL WORLD FAIR

EVERY FEW YEARS THE MINI WAS IMPROVED, AND WHEN THE 1970S FINALLY ARRIVED, SO DID THE CONTROVERSIAL CLUBMAN CLAN, HELPING SALES CLIMB EVER HIGHER. CELEBRITIES, MOVIEMAKERS, AND ICONOCLASTS MADE THE LITTLE CARS COOL.

The Mini's bigger sibling, the Morris 1100 (there was an Austin equivalent, of course), took Issigonis's packaging concept into the family car sector and introduced fluid-filled Hydrolastic suspension.

The Mini, over its first five years, was a small car that made the entire global motor industry stop and ponder its future direction. From a hesitant start, as teething problems were quickly fixed, Mini sales gathered enormous momentum. Helped by the proliferation of different models, and the Cooper image boost—on race tracks, roaring to rally wins, and zipping between trendy music venues and alluring new boutiques on London's most stylish streets—the million-car milestone was reached in 1965. Alec Issigonis was beaming as he drove it off the Longbridge assembly line. For a huge number of drivers it was the most logical yet also the most fashionable choice, which was an admirable achievement for BMC at a time when car ownership, with almost all hire-purchase limits now cast aside, was exploding.

By the end of 1962, the Mini had a big brother: the Austin or Morris 1100. This four-door family saloon was the next step in Issigonis's plan, using a similar front-drive/transverse layout but with his next-generation of Hydrolastic suspension system for a more forgiving ride while still offering safe and nimble handling. At around the same time, BMC acknowledged the word on the street and rebranded the Austin Se7en as, of course, the Austin Mini; meanwhile, after car tax was slashed by 10 percent in the budget, the cheapest Mini fell in price to £495, which made it even less expensive than when it had launched in 1959.

Issigonis's ego, which was rarely hidden anyway, received a major boost in 1964 when the first rival car to follow his Mini/1100 transverse-engined, front-wheel-drive template was announced. This was the Autobianchi Primula, made by a Fiat

This obscure car is the Autobianchi Primula, unveiled by Fiat in 1964; quite apart from being one of the world's earliest production hatchbacks, it was the first rival product to adopt the Issigonis front-drive/transverse-engine layout.

subsidiary, although the small family saloon differed in having its gearbox mounted end-on to the engine, rather than in the oil sump below it. The Primula is largely forgotten today, but it paved the way for the extremely successful 128 and 127 models that sustained Fiat's fortunes right through to the 1980s. Throughout that time many more cars would adopt the Mini's pioneering concept for efficient packaging and largely foolproof roadholding.

One area, though, where rivals didn't follow BMC was in suspension. And rather than join the herd and use coil springs, the Mini underwent yet another radical move in 1964 when it switched from its unique rubber cones to Hydrolastic like the 1100.

This was an interconnected system incorporating sprung displacer units, which used rubber valves to push fluids between the front and back wheels. It was inspired by Citroën's self-levelling systems and was part of Issigonis's zeal to harness the very latest technology to enhance comfort. Each wheel carried a cylinder about the size of a 1-liter oil can containing the spring and damper in one unit and using a frost-resistant water emulsion as the damper fluid. The hydraulic chambers on the front and rear wheel dampers were connected by pressure hoses on each side of the car.

-O [left] A Mini 1000 Mark II in 1968 in its natural habitat, London's busy streetscape; bumper overriders were standard on the Super De Luxe.

[below] The Morris Mini Super De Luxe, in Mark II form, could now be had with either 848 or 998cc A-Series engine options.

The full spectrum of the Mark II range at the point, in 1968, when the British Motor Corporation became British Leyland . . . and the Mini one of its least problematic cornerstones.

That meant that whenever a front wheel hit a road bump, some front hydraulic fluid was forced into its rear counterpart, also lifting up the body slightly at the rear. And vice versa, of course.

The ride quality was better, for sure, less jarring, although both engineers inside the company and customers out on the road came to the conclusion that the sharpness of the Mini's handling suffered slightly and that the overall improvement was negligible. The former rubber cone system had provided such a terrific trade-off between instant steering response and absorption of most road-surface imperfections that it was very hard to provide significant improvement.

A Mini refinement that really was worthwhile, for some drivers anyway, was the 1965 introduction of automatic transmission—a clever, compact design from Automotive Products that beat its very few rivals among small cars in having four speeds over the conventional three. By then, too, some dashboard controls had been rearranged so that they were finally all within reach of a driver wearing a seatbelt!

At the height of the Mini's international fame, as it took the third of its Monte Carlo Rally victories in 1967, BMC seized the moment to revamp the whole range and announced it officially as the Mark II range. The principal mechanical change was that the spritelier 998cc engine option was now offered in the standard-shape car, van, and pickup, as well as the Elf/Hornet, with more torque—38 pounds-feet, up from 32—and 4 more brake horsepower to take output to 38. And all models

benefited from a turning circle reduced from 32 to 28 feet, which made every Mini that bit nippier. The Mark II cars were also refreshed stylistically: there was an enlarged rear window, much larger rear light clusters, and a square-cut radiator grille shape. Inside, redesigned seats and much improved trim upped comfort levels.

Orders for the cars as they steadily evolved into what most people assumed would be middle age just got bigger and bigger, and as 1968 turned into 1969, annual production topped the quarter-million mark for the first time. This reflected the Mini's tiny running costs, excellent fuel economy, and general trustworthiness at a time when the British economy was dogged by higher inflation and unemployment, giving rise to freezes on wages and prices, credit curbs, and interest-rate hikes. The Mini thus became a cornerstone of British Leyland, formed by a government-brokered merger of BMC/Jaguar, brought together two years earlier as British Motor Holdings, and Leyland Motors, which as well as being a major truck manufacturer also owned Triumph and Rover.

However, the tumultuous corporate changes, starting in May 1968, shone a harsh new light on how the Mini and other Issigonis-inspired cars were regarded. The man in charge of the gigantic new company, which was the world's third-largest carmaker with 198,000 employees and 48 factories—dwarfing Toyota and Volkswagen—was Sir Donald Stokes. Under his tenure, Leyland had become a highly successful business, selling lorries and upmarket cars. By contrast, BMC's high-volume/low-margin approach refused stubbornly to yield returns; for 1968, for example, it made a £3 million loss on sales of £467 million. Stokes (who became Lord Stokes in 1969) was now in charge and determined to shake things up, and the Mini was in his sights for some must-do-better treatment. The full extent of the changes was revealed in October 1969, and for Mini purists they came as a shock.

The most striking technical change for the Mark III range was in its suspension. Hydrolastic, which was expensive to manufacture and fairly unpopular with owners because of its maintenance costs, was abruptly axed on the standard 850 and 1000 saloon and the original rubber cone system reinstated.

[top left] Inside the Mark II Mini in 1968, where the seats had been designed and the trim, if you can believe it from this picture, had been upgraded.

[top right] A Cooper S zips round Parliament Square in the shadow of the Big Ben clock tower in 1968; like all the Mark IIs, it had a new, squared-up grille shape and larger back window.

No one likes a parking ticket, although this Mark II–owning couple is laughing it off, but yellow lines were proliferating in the 1960s, putting parking space a premium and so intensifying the Mini's urban appeal.

It was unacceptable, said the new Leyland-biased management, for a mainstream car to suffer the cut-price indignity of sliding windows in the 1970s, so the Mini's doors were redesigned to take wind-down ones and also to conceal the utilitarian, previously external hinges. To accommodate these changes, Issigonis's treasured door bins had to be sacrificed.

The Moke had already been dropped from the UK lineup, dispatched for its sins to live on in Australia. Now the Wolseley Hornet and Riley Elf (indeed, the whole Riley marque) got the chop, and the new luxury Mini was announced as the 998cc Clubman. Designers recruited from Ford gave the car a new, Ford Cortina–like square nose treatment that added 4.3 inches to the front of the car. And rather bland and anonymous it looked too. All other dimensions were unchanged, but the Clubman had a more inviting, better-quality interior with an instrument cluster directly in front of the driver. Replacing the Traveller and Countryman was a Clubman estate, with real wood trim replaced by wide strips of Di-Noc fake wood along the flanks of the car.

Simultaneously with these changes, Mini production at the old Morris plant at Cowley ended, with all cars now emanating from the Austin factory at Longbridge.

CONTINUED ON PAGE 88

THE REAL ITALIAN JOB . . .

The Mini was a hugely successful car for Britain, but its popularity was in no way restricted to its home market. There were one or two markets where it didn't catch on, notably the United States, but almost everywhere else the car was sold it proved desirable and, mostly, robust once the early manufacturing design faults had been ironed out. By the time that the two millionth example had been built in 1969, it was revealed that 43 percent of Minis built so far had been exported, and almost half the shipments had been of completely knocked-down (CKD) kits for local assembly. No other purely British car came anywhere close for global success.

The export department at Longbridge throughout the 1960s and into the early 1970s was a frenetically busy place, as Minis in contrasting states of completion were dispatched all over the world. There were assembly plants in Australia, New Zealand, and South Africa, which, thanks to tax rules there, needed to meet different quotas for local content levels. In 1975, 66 percent of a Mini "made" in South Africa was deemed local, in value terms. This meant the quantity of Mini components manufactured at each factory varied hugely; some made their own bodies and engines, while others merely bolted and welded the imported sections together onsite. But most versions of the Mini on sale in these far-flung outposts had numerous small differences from their cousins back home. Australia, for instance, adapted the cars to include winding windows in 1965, four years before the UK, to better meet local buyers' demands.

By the early 1970s, British Leyland was also manufacturing Minis in Spain and Belgium, before there was a single European market, which helped somewhat to counter diminishing sales in the old commonwealth territories as the Mini aged and competition from small Japanese cars started to bite.

One country that had its very own Mini ecosystem was Italy. There, the maker of the world-famous Lambretta scooter, Innocenti, had first built British cars in 1960 after agreeing a license deal with BMC to manufacture the Austin A40 for the Italian market. It soon added its own,

[above] A late-model Innocenti Mini Cooper 1300 looking lush with its Rostyle wheels, two-tone paintwork, unique grille, and even antilift aerofoils on its windshield wipers.

[opposite top] Inside Innocenti's bustling Lambrate assembly plant in Milan, sometime in 1968, with Minis already dominating the production lines.

[opposite bottom] When British Leyland refused to update the Mini, Innocenti had a go itself with the three-door Mini 90, calling in Bertone to look after the styling.

restyled version of the Austin-Healey Sprite and then the Austin 1100, which was sold as the Innocenti IM3. The Innocenti Mini-Minor joined the lineup at the Milan factory in 1965, and once a Mini Cooper and Mini T (Traveller) were added a year later, sales of the Italian-made Minis soon outstripped all Innocenti's other four-wheeled products and gave the company about 5 percent of the Italian market, which was comprehensively dominated by Fiat. The Mini-Minor 850 was always more expensive than its Fiat equivalent, the 850 saloon.

Innocenti Minis started as kits of parts supplied from Britain, but there was always a high degree of Italian-sourced ingredients, including many of the body panels, tires, radiator, heater, wheels, and glass. The specifications of the cars evolved at a slightly different pace from their British equivalents, but they were generally much better finished and had some lovely features, such as a unique five-dial dashboard in the Cooper. From 1970, shortly before the Cooper option became the 1,275cc Cooper S–matching Innocenti Mini-Cooper 1300, swiveling front quarter light windows were added across the range—which pleased Italian drivers no end for those baking-hot summer drives—in stainless steel frames, which made the cars look even smarter. They were exported to several European countries, Germany and France included, and the sought-after Cooper name survived because the licensing agreement was separate from the curtailed British one. In fact, British Leyland did briefly own Innocenti's car-making operation, but Mini manufacture ended in 1975 after the company introduced its 90 and 120 small hatchback range. After almost fifteen years, Innocenti's management was desperate for something more modern to sell to style-conscious Italian customers and, when nothing was forthcoming from Britain, created this new car themselves, using trusted Mini subframes and power units.

In ten years, Innocenti produced 440,000 Minis. This production contributed massively to the car's popularity on mainland Europe while being largely ignored and unsaluted by the head office back in Birmingham . . .

[above] It looks pretty groovy now, of course, but the Cortina-like snout of the 1275GT and Clubman was perplexing to Mini purists in 1969. Additionally, go-faster stripes were new on the sporty Mini—the Coopers didn't really need them—and the 1275GT headed into the 1970s with the sort of garish livery that was then all the rage.

[right] The 1275GT was the Clubman-based heir to the Mini Cooper performance crown, although the 59-brake-horsepower engine didn't have much in the way of exotic componentry, and the alloy wheels were fakes.

Yet neither of these marque names was to be seen any longer on the new lineup: Mini finally became a make of car in its own right. It made total sense, of course, because hardly anyone uttered the full brand name, but traditionalists couldn't help but feel the new corporate leviathan of British Leyland was chipping away at what was now the Mini's pukka heritage.

Hydrolastic suspension remained in use for the new Mini Clubman saloon and estate, and also lingered on in the Mark III Cooper S too. Like the rest of the 1969-vintage Minis it gained wind-up windows, but to trim manufacturing costs the contrasting-color roof panel was dropped. The 1,275cc Cooper S lasted only until 1971, when even more penny-pinching saw British Leyland scrap its royalty deal with John Cooper to save £2 on every car. This decision by Lord Stokes has been lambasted as short-sighted and petty because it junked so much of the Mini's goodwill and image, but the context is often forgotten. The Cooper Formula 1 team had been in slow decline and absent from the tracks since 1969.

For those who loved spirited driving, though, all was not lost. The new performance Mini in the 1969 shake-up was the Clubman-based 1275GT. With a 59-brake-horsepower, 1,275cc A-Series engine, albeit depending on just a single carburetor (British Leyland offered a conversion kit to add another if you wanted that), it gave a fairly vivid account of itself, with 0 to 60 miles per hour in 14.7 seconds and a top speed of 86 miles per hour. The boy-racer exterior get-up included go-faster

CONTINUED ON PAGE 92

[top left] A Clubman photographed in Austria with its stylish owner (probably); the new model certainly helped sustain the Mini's popularity on mainland Europe in the early 1970s.

[top right] Cliff Richard starred in a 1973 movie set in Birmingham called *Take Me High*. His character, Tim Matthews, drives a customized Mini Clubman so fully equipped that it even has a built-in electric razor to maintain Cliff's wholesome, clean-shaven looks.

[above] The Clubman interior had all-new seats and trim in ribbed, padded vinyl, which was considerably more comfortable than before, plus wind-down windows.

THE ITALIAN JOB

Time and again, *The Italian Job* has been voted one of the greatest British films of all time. At least, it certainly is in the minds of the British public, which took the gold-bullion-heist comedy caper to its heart as soon as the film opened in 1969. Star Michael Caine's withering line "You're only supposed to blow the bloody doors off!" and the chirpy lyrics to Quincy Jones's "Getta Bloomin' Move On! (The Self Preservation Society)" have entered the British English lexicon and are frequently heard in banter across the land from offices to pubs and chanted at football matches.

The story revolves around criminal gang leader Charlie Croker (Caine), who is released from prison and is immediately drawn into an ambitious plan to steal the precious metal as it's delivered from China to Italy, outwitting the Italian mafia, police, and security services. Noël Coward plays Mr. Bridger, the underworld power broker who approves Croker's harebrained scheme on the grounds that Britain needs the money. The actual robbery involves three Mini Coopers for the final getaway, chosen, of course, for the cars' phenomenal agility and small size.

Created in a world long before computer-generated imagery (CGI), the film used expert stunt drivers, led by Rémy Julienne, to drive the cars through a traffic-clogged Turin after the city's traffic-light system has been paralyzed by Croker's tame computer boffin Professor Peach (Benny Hill). In scenes that many find unforgettable, the Minis are driven in and out of subways, through sewers, across the roofs of buildings, over frothing levees, and finally into a moving bus before each bites the dust as the gang escapes. The ending includes the ultimate movie cliffhanger.

The film was directed with verve by Peter Collinson and produced by Michael Deeley, who turned down a lucrative offer from Fiat to have its cars used in the getaway because the all-British Mini was so central to the essence of the script. Hence, the Mini Cooper S enjoyed a truly starring role—at the very height of its global fame—and its effervescent dynamics, with just a dash of movie magic, were on full display.

"*The Italian Job* would be the longest commercial for a car ever made," said the producer Michael Deeley in his 2008 autobiography. Nevertheless, he found BMC/British Leyland had a strange apathy to the living legend in its care; the company only agreed to supply him with six cars "at cost," meaning the film's budget had to stretch to accommodate another thirty at full retail prices. "Many of those who worked on the picture felt BMC's attitude was a sad reflection of the British car industry's marketing skills," recalled Deeley. Nevertheless, the sun-kissed fantasy, with Minis at its center, has proved enormously powerful to sustaining the Mini's popularity for decades afterwards.

Two of the Mini Coopers from *The Italian Job* about to descend the steps into one of Turin's subways, as footie fan extras look on in wonder.

[left] One of the showpiece stunts in *The Italian Job* was the sequence of a Mini leaping from one building to another, quite a feat for Rémy Julienne and his team of stunt drivers.

[below] *Italian Job* Minis being pursued by the city police (in an Alfa Romeo Giulia) during a scene shot on the rooftop test track of Fiat's Turin factory.

body stripes and Rostyle pretend-alloy wheels, while inside the instrument cluster included a rev counter.

There were a couple of uncomfortable issues surrounding the Clubman and 1275GT. Firstly, the restyling did nothing to improve the car's functionality. Under the modern look of the new surfaces was the same old car, now a decade old and so, by normal motor-industry standards, overripe for replacement. And secondly, the Clubman transformation seemed to focus on removing the Mini's cheeky, snub-nosed charm—its classless appeal was under siege from the need to add a sort of corporate aspiration factor to the car. Little wonder, then, that the era when the Mini was the darling of Britain's showbiz royalty was fading out.

The cars had been owned by everyone from ballet legend Margot Fonteyn to Hollywood heartthrob Steve McQueen, from fashion icon Twiggy to comedy eccentric Spike Milligan and pop star Spencer Davis. All four of the Beatles owned Minis. Paul McCartney and Ringo Starr ordered their own specially customized Coopers. John Lennon bought one in 1964 even though he didn't have a driving license; he gave it a distinctive, psychedelic paint job. George Harrison lent his Mini to Eric Clapton in 1967 and only got it back three years later. They were the small cars to own and be seen in for all the beautiful people in London and well beyond—Enzo Ferrari was a noted Cooper devotee, and actor James Garner was a frequent sight around L.A. in his.

[top] One of the very last Mark II cars, a Morris Mini Super De Luxe, still clinging to those sliding windows that were a throwback to 1950s austerity.

[above] The 1,275cc Cooper S lingered on in the Mark III range into the 1970s, minus its contrastingly painted roof, until Lord Stokes decided to axe royalty payments for using the Cooper name, and a special era was over.

[left] Austria again, and a Mini 850 is causing a flurry of interest in a marketplace—or, at least, that could be what the charmer from Salzburg on the left is interested in.

[below] The Mark III Mini van of 1969 stuck with sliding windows in line with its utilitarian purpose.

CHAPTER

EVERGREEN OR SIMPLY NEGLECTED?

IF THE LIFE AND TIMES OF THE MINI IN THE 1970S ARE DEFINED BY ANYTHING, IT'S A CURIOUS STAGNATION. THE CAR MAY HAVE ALREADY ATTAINED THE STATUS OF AN AUTOMOTIVE NATIONAL TREASURE, BUT RATHER THAN RESPECT, IT SEEMED THE MINI SUFFERED FROM APATHY.

Not the ideal car for long motorway journeys, but the Mini 850 of the mid-1970s would almost certainly get you there.

In 1971, the Mini Clubman and 1275GT joined their round-nosed stablemates to complete the total reversion to dry rubber cone suspension. Then, in 1973, a much-improved rod-shift gear change was introduced across the board, ridding the standard Mini of the vague "magic wand" that had been a feature from the start.

The sporty 1275GT became the first Mini with 12-inch wheels in 1974, along with bigger brake discs. Alongside those was an option that brought a little innovation back to the Mini: Dunlop's Denovo wheels and run-flat tire system. These became standard equipment in 1977 and would remain part of the package until the 1275GT was dropped in 1980. Meanwhile, in 1976, a barely trumpeted Mark IV range came with a small basket of trim and finish refinements—two-speed windshield wipers and electric washers the most worthwhile part and the gradual introduction of matte-black painted radiator grilles the most obvious. One way or another, the range throughout the decade always offered an 850 or 1000 saloon, the Clubman saloon and estate, the 1275GT, and the commercial van and pickup pair.

The response from competitors to the Mini's success in the 1960s was fascinating. While Germany's even more venerable Volkswagen Beetle swelled hugely in popularity, in France the pace was made by the 1961 Renault 4, which offered front-wheel drive (with a longitudinally mounted engine) in a compact, softly sprung five-door estate car. In Italy, the Fiat 850 of 1964 stuck resolutely to a rear-engined layout, but this technical format had done little to help the Hillman Imp from Britain, launched the year before; this quirky four-seater with its rear window opening as a kind of

early hatchback suffered design and quality problems and posed little threat to the Mini's small-car dominance.

No other company copied the Mini directly, but by the 1970s many mainstream manufacturers were readying cars that took the Mini's best aspects and put them into a more versatile, modern package. Not for nothing were they nicknamed superminis. First came Datsun with the Cherry and Fiat with the 127 in 1971, both with transverse engines and front-wheel drive, but the Renault 5 (front-wheel drive/longitudinal engine) a year later really set the pace because it had folding back seats and a hatchback third door that reached right down to bumper level. Then, in 1976, the Ford Fiesta (shown on page 95) combined front drive, a transverse engine, a full-depth hatch, and neat, modern styling to absolutely define the new supermini breed. It was a massive hit right from the word *go*.

[top] With so many millions of Minis on the planet, examples that had failed Ministry of Transport (MoT) testing or been damaged in a crash tended to be worthless in the 1970s, with wrecks often abandoned in scruffy corners of Britain's towns and cities.

[above] Datsun was the first of Japan's manufacturers to join the transverse-engined, front-wheel-drive club with the 1971 Cherry 100A.

Back in 1971, the Mini had enjoyed its best ever year on sale, with an astonishing 318,475 built and 102,000 of those sold in the United Kingdom. On October 25, 1972, the three millionth was built, and in 1976 the four millionth. By 1977, though, after the Fiesta's first full year on sale, Mini manufacture had slumped by a third, and by 1980 it had more than halved.

By then the Austin Mini Metro had arrived as British Leyland's entry into the frothing supermini market sector. Bearing in mind that Issigonis had started work eagerly on a Mini replacement, code-named 9X, in 1966, it was at least ten years late. There were numerous reasons why the company hadn't gotten its act together sooner. One was the need to pour resources into the Morris Marina and Austin Allegro to compete with the best-selling family cars from Ford, namely the Escort and

Celebrating the Mini's first fifteen years at the Longbridge factory in 1974. The raised bumpers indicate this example was bound for Canada, where the Mini was a very popular small-car import.

Cortina; another was an ongoing internal argument about what specific size a Mini replacement should be. But set against these troubling issues were the crumbling financial viability of British Leyland brewing up before its bankruptcy and nationalization in 1974–1975 and the Mini's uncanny role as the perfect car for the turbulent times of the 1970s, with its economic squeeze, industrial unrest, and fuel crisis. The Mini was cheap to buy and run, and, just as in the mid- to late 1950s, many people still found it was just the ticket for everyday motoring. It was the kind of car you wanted to own when, for instance, gas prices doubled in the United Kingdom between January 1974 and January 1975 as inflation sprinted ahead. The fact that actually making the car produced very little profit for the inefficient British Leyland Motor Corporation behemoth was, for consumers, of little consequence. After all, in 1973, taking inflation into account, a Mini 850 at £739 was cheaper than ever; that would have been £370 adjusted to 1959 prices.

By the late 1970s, Minis were serving as the most popular first cars of choice to a second generation of newly qualified drivers. The whining gearbox, the slightly awkward driving position that was more delivery truck than sports car, quite possibly a

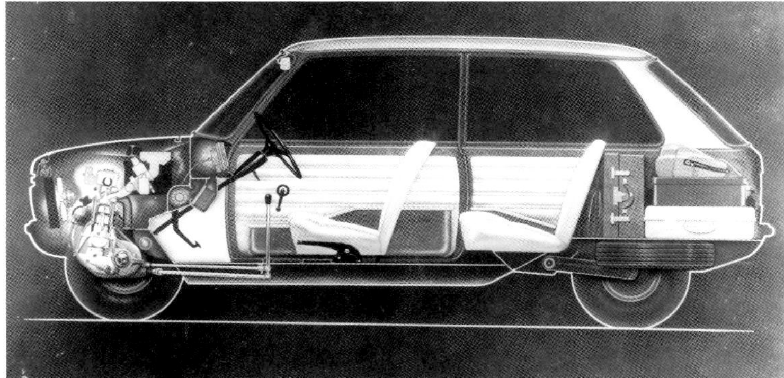

[top] Tens of thousands of rookie British motorists could throw away their L-plates after learning to drive in a Mini in the 1970s and 1980s.

[above] This hatchback design proposal from Alec Issigonis for a Mini replacement, code-named 9X, failed to progress beyond a running prototype, which was nevertheless impressive.

clothes-peg gripping the feeble choke to hold it out on cold mornings, and of course the exhilarating, limpet-like cornering—these were all things that tens of thousands of new drivers thought were totally normal on every car. And, indeed, for anyone buying a new Metro hatchback after its much-vaunted launch in 1980—through advertising and news reports, the entire nation was told that it was a savior of the British motor industry—many of the Mini's endearing characteristics remained familiar. This was because, in order to make the new three-door hatchback viable, Mini-style subframes carrying the A-Series engine, transmission, and new Hydragas suspension units were used on the Metro. While the Metro boasted a completely new hatchback structure, many of the Mini's traits lived on within it. Great when it came to on-road agility, but not so good for cruising on motorways with the tiresome burden of a thrashy four-speed gearbox.

The boxy Metro, though, wasn't a replacement (even though it did kill off the Clubman series). The traditional Mini carried on alongside it at Longbridge, and from 1980 in a revised form that was quietly impressive. This was down to a comprehensive project to fit sound-deadening material all over the car—such as in the roof, around the exhaust tunnel, and across the front bulkhead—that made it vastly quieter inside. The 848cc engine was dropped and a new A-Plus 998cc engine soon standardized on the stripped-down City E and comfort-orientated HLE, with a raised compression ratio and revised carburetor to give a leaner mixture, plus a higher final drive ratio and special Dunlop low-rolling-resistance tires. The result was an 86-mile-per-hour car that was pretty lively, with 0 to 60 miles per hour in 17.5 seconds, but very miserly on fuel, with over 60 miles per gallon easily attainable at a constant 50 miles per hour and a 40-mile-per-gallon average. Not long afterwards came the 1982 Mayfair model, with plush Raschelle upholstery and matching door trim, and the Mini-first fitment of front-seat head restraints. Along with its tinted glass and optional wide alloy wheels with black plastic wheel-arch extensions, this was a very belated attempt to bottle the custom-built luxury character of 1960s cars built for celebs and sports stars.

In its twenty-fifth birthday year, superb new superminis like the Opel Corsa/Vauxhall Nova, Peugeot 205, and Renault 5 Mark II were hitting the road and making the original Mini seem decidedly decrepit. If the charismatic little Brit was going

CONTINUED ON PAGE 104

[top] The Austin Mini Metro took its bow in 1980 as British Leyland's belated supermini. It didn't actually replace the Mini, but it did use subframes similar to its illustrious running mate.

[above] Minis with six wheels were often showily designed kit cars, but a few were converted from standard vehicles, such as this elongated, eye-catching pickup.

[top] The Mini range for 1981, after the arrival of the new Metro. From left to right, the 1,000HL (formerly the Clubman), estate 1,000HL, and 850 City.

[above] A 1960s Mini van amusingly shortened into a stumpy two-seater roadster can park end-on to the curb in a 1970s British high street.

[above] Images released by BMW in 2013 emphasized the Mini's classic credentials in the company of another venerable classic, the Porsche 911. A great bit of color coordination highlighted by the overcast weather.

LIMITED EDITIONS

The idea for a sales-boosting short-run Mini was first tried, tentatively, in 1976 with the Mini 1000 Special Edition. Its Brooklands green paintwork, gold coachline, tinted glass, and orange brushed-nylon upholstery were certainly eye-catching, and all three thousand cars sold quickly at £1,406 apiece. Emboldened by the idea, and looking for something to celebrate the little car's twentieth birthday, British Leyland next issued the 1100 Special in 1979. Technically speaking, it's an interesting car, being the only UK-built standard-shape Mini ever with a 1,098cc engine. The liveliness was emphasized by a choice of metallic rose or silver main body, with contrasting tan or black vinyl roof respectively, side stripes, fat alloy wheels, extended wheel arches, tartan seats, and 1275GT-style sports steering wheel and instruments. It proved so popular that 5,100 were sold against the originally intended 2,500 run. The 1983 Sprite, meanwhile, added the alloy wheels/wide arches touch to a Mini City, with new decals and herringbone-pattern seat facings.

That was it for the 10-inch-wheel cars, but soon after the 12-inch wheel range arrived in 1984, along came the Mini 25 as the car motored past its silver jubilee in silver-painted, red-accented style. As ever, the whole run of five thousand cars was snapped up in double-quick time by fans, and, from then until the entirely predictable Mini Thirty edition in 1989, the lineup rarely lacked a special-edition member. The Ritz, Chelsea, Piccadilly, and Park Lane all celebrated the Mini in its natural London environment, while others, including the Flame, Red Hot, Jet Black, and Advantage, were all exuberant presentations of the basic car resplendent with their individual liveries, paint jobs, decals, wheels, and myriad small luxury touches.

Perhaps the most significant from a cultural perspective was the Mini Designer of 1988, created in collaboration with British fashion legend Mary Quant, who had been central to the popularizing of the miniskirt in London in the 1960s. Quant had bought a brand-new Mini the moment she passed her driving test; now she added bold black and white stripes to the upholstery of what she called "a really fashionable, happy and smiling car."

This 1100 Special of 1979 celebrated the Mini's twentieth birthday with metallic silver or rose paintwork and is significant for being the only standard-shape Mini to run the 1,098cc A-series engine.

[top left] Cherry-red paintwork (black was offered too) along with chrome grille and bumpers look great on the Thirty, which marked the Mini's third decade on sale.

[top right] Mary Quant herself in the limited-edition Mini she created, complete with striking black-and-white upholstery; a Mini was the first car for the designer so closely associated with the miniskirt and other defining 1960s styles.

[above left] Silver bodywork, a gray and red coachline, and a very plush interior made up the livery for the Mini 25, commemorating the car's quarter century in style, and on 12-inch wheels, in 1984.

[above right] Minilite-style alloy wheels and British racing green paintwork were partnered with a full-length, electrically operated fabric sunroof on the 1992 British Open Classic edition.

It certainly ate up creative ideas, but the production line at Longbridge was kept humming throughout the late 1980s and 1990s. The editions themselves could be pretty large, such as the 3,725-off Ritz of 1985, or quite small, as in the 500 Mini Tahitis offered un 1993; they could be straightforward paintwork-and-stickers makeovers, like the 1990 Mini Studio 2, or minor works of the customizer's art; the 1992 British Open Classic (1,000 made) had a full-length electric sunroof, and the sought-after Paul Smith edition in 1998 (300 for the United Kingdom) featured a body-color dashboard in blue, beautiful leather seats, a denim head-lining, and contrasting citrus-green underhood, fuel tank, and trunk floor.

to keep going at its modest but feasible production rate of 30,000 to 40,000 cars a year, it had to meet some minimum standards. Hence, in 1984, 12-inch wheels were standardized across the surviving City and Mayfair models to accommodate more effective 8.4-inch front disc brakes, and plastic extended wheel arches were fitted to house them. Then, in 1989, when the new renamed Rover Group could hold out against legislation no longer, the minimum was done to keep the Mini road legal by modifying it to run on unleaded gasoline.

An interesting new source of sales success was found in Japan, where the dainty Mini had been developing something of a cult following since the mid-1970s. Austin Rover Japan took over from a private Mini importer in 1985 and started selling the Mini there as part of its official range. Within five years annual sales rocketed from 1,500 to 11,000, propelled by new factory-offered options such as air conditioning and automatic transmission. In addition, buyers were spending an average of £500 on additional accessories, like a replica of the roof-mounted spare-wheel rack fitted to the 1967 Monte Carlo–winning Mini Cooper S.

"We were fortunate that we surfed a wave of interest in imported cars at the time of Japan's 'bubble economy,'" said David Blume, who ran Austin Rover Japan at the time. "For Japanese people who love cars, owning [a Mini] is the fulfilment of a

[top left] Welding the Mini together was still a job largely done by hand in the mid-1980s, just as it had been in the far-back 1950s when the car was conceived.

[top right] The 1985 Mini City, now riding on 12-inch wheels and boasting front disc brakes for the first time on a bargain-basement model.

[bottom left] Aston Martin Tickford built this fully loaded luxury one-off for an individual client in the mid-1980s. Incorporating the rectangular headlight stacks was undoubtedly a coachbuilding achievement, even if the finished result looked a mite strange.

[bottom right] The 1982 Mayfair was a regular production model, boasting velour seats and the Mini's first ever standard headrests; alloy wheels and black wheel-arch extensions were optional extras.

June 1980 saw the introduction of the new quiet Mini, after the little car was subjected to a program of sound deadening to keep it just about abreast with other modern cars in this capacity. It was surprisingly effective.

Rowan Atkinson, fully in character here as Mr. Bean, driving the 1970s Mini that featured in the comedy show of the same name; the Mini might have been mocked, but it gained renewed exposure all over the world.

dream. And also, of course, in a crowded country like this, it still delivers its promise: it still gets four people—okay, not with much luggage—around the city, and then fits into the smallest of parking spaces. It's the Tamagotchi factor: being cute and physically small is the appeal."

The five-millionth Mini was built in 1986 and driven off the production line by car-loving DJ and TV presenter Noel Edmonds. It was rather wonderful that Sir Alec Issigonis could witness his creation become—by a very long way—the biggest-selling British car of all time; he passed away soon afterwards, in 1988. In topping five million, the Mini took its place at the select top table of motoring legends to have surpassed that mighty total, jostling for parking space with the VW Beetle and Golf Mark I, Ford Model T, and Fiat 124/Lada 2141.

CHAPTER

THE COOPER REVS UP AGAIN FOR THE 1990s

TAXI
V205 LOE

THAT THE MINI HAD SOMEHOW MANAGED TO EVADE THE AXE THROUGHOUT THE 1970S AND 1980S IS, IN RETROSPECT, ONE OF ITS MOST INCREDIBLE UNSUNG ACHIEVEMENTS. NOW, IN THE LAST DECADE OF THE TWENTIETH CENTURY, IT WAS BOOSTED BY THE RETURN OF A FAMOUS NAME.

The first of the revived Coopers was this 1,650-car limited-run model in 1990 with signed John Cooper hood stripes, sunroof, and spotlamps; the engine was the MG Metro's 61-brake-horsepower single-carb unit.

In 1990, annual Mini production was about one-sixth of its 1971 peak, at about 46,000. UK sales at some 10,000 were 90 percent less than the same highpoint. The Metro, the Mini's natural successor, had just seen a major revamp after its ten years on sale, with brand-new engines, new interconnected Hydragas suspension, and a substantial restyle inside and out. The most newsworthy thing about the Mini that year was the option of a catalytic converter to clean up its exhaust emissions.

The popularity of the Mini in export markets, especially Japan, was one factor that kept the Birmingham production line rattling along. Another was that the Mini's manufacturer had plenty to deal with on other fronts, while the Mini itself offered little in the way of extra strife. Changes of name reflected the desire to crawl upmarket, with BL Cars becoming Austin Rover Group in 1982 and Rover Group in 1986. Similarly, brands that didn't align with this strategy were unceremoniously dumped, meaning the end for Austin, Morris, and Triumph.

Underperforming models were sidelined, in particular the mediocre Maestro and Montego, as the company's new Rover models were jointly developed with Honda. Jaguar, Leyland Trucks, and Unipart had all been separated out from the company and sold off. And then there was a change of owner. In selling the company to British Aerospace in 1988, the British government ended Rover's state custodianship after thirteen turmoil-filled years.

However, the Mini also remained one of the best-loved single cars ever made—present owners still relished driving theirs, and millions more had golden memories

John Cooper pictured in 1994 with his Grand Prix–edition Mini Cooper, issued to commemorate the thirty-fifth anniversary of the Cooper team winning the Formula 1 World Championship.

of the days when they used to do the same. Indeed, there was a brief and bizarre attempt to pair the Mini's legendary roller-skate fun factor with the go-faster gizmo of the day in 1989 in the Mini ERA Turbo, with its tacky, body-kitted appearance as questionable as its engineering. Yet the magic name Mini Cooper still resonated loudly. It was a fully paid-up part of the English language, in the United Kingdom certainly.

John Cooper remained well aware of the magic pull of his surname, and in 1982 he tried to get an Austin Metro Cooper off the ground. As this would have clashed head-on with the MG Metro, he was prevailed upon to abandon the project, but dealers and customers in Japan latched on to the idea of a revived Mini Cooper. John Cooper Garages was asked to produce an authentic Cooper tuning kit for the 998cc standard Mini sold through Mini Maruyama in Tokyo. Tuning firm Janspeed collaborated on the conversion, and it was an instant hit, selling over a thousand examples in the first year. John Cooper decided to launch it in the United Kingdom

and, after just one magazine article, immediately received seventy orders for the £995 package.

The clamor caught Rover by surprise, and it swiftly decided to relaunch the car. It was fairly easy to install the 61-brake-horsepower, 1,275cc, single-carburetor MG Metro engine into the Mini, add an oil cooler, and add a three-branch exhaust manifold with twin downpipes and silencers. The only tricky part was engineering an additional electric cooling fan so the new Cooper could meet recently introduced drive-by noise rules. The Minilite-style alloy wheels, black plastic wheel-arch extensions, and contrasting white roof looked fantastic, while the rest of the interior trim stuck closely to the luxury feel of the recent Mini Thirty limited edition.

The car was launched in June 1990 at £6,995, and the timing for the factory Rover Mini Cooper revival proved absolutely spot-on. There were 1,650 examples of the launch edition (1,000 for the United Kingdom, 650 for Japan) with hood stripes carrying John Cooper's signature, a glass sunroof, and twin spotlamps, and they sold out in a few days. Little wonder that a regular production model was unveiled four months later, minus the oil cooler, hood stripes, and sunroof, but £400 cheaper. Within a month, nearly one in three Minis sold was a Cooper, and John Cooper developed a further performance kit for an officially sanctioned—but strictly

[top] A badge declaring "1.3i" and a new winged hood emblem denoted this as the 1991-on fuel-injected Mini with a catalytic converter as carbs were consigned to history.

[above] The regular production 1990 Cooper did without such niceties as an oil cooler but was £400 cheaper than the introductory special, at £6,595. The Mini Cooper's timeless appeal flourished in the 1990s. Buyers loved its authenticity, and sales were amazingly buoyant considering its advancing years.

Japanese demand was one key factor keeping the original Mini alive; here new Coopers embark on a several-week sea journey in about 1994 to meet new owners in Tokyo and Osaka eagerly awaiting their arrival.

aftermarket, to avoid new car-build regulations—Cooper S conversion at £1,751, offering a jump up to 78 brake horsepower and the options of adjustable shock absorbers and low-profile tires available to sharpen up the roadholding.

There was no leap in overall Mini sales, but customers were paying more per car, and the Mini Cooper revival brought a vast amount of goodwill to Rover for very little outlay. Indeed, it put the company in the vanguard of retro revivals. Within a couple of years, Rover had also revived the MGB; later in the 1990s, Volkswagen would launch a new Beetle and American giant Chrysler would give its PT Cruiser family car a strongly 1950s custom-car character.

Only a year after the Cooper made its big rebound, the carburetor started to slip out of the Mini's circle, as the Cooper's engine switched to single-point fuel injection with catalytic converter as standard. By May 1992, the 50-brake-horsepower motors in the Mayfair and the new base-model Sprite, which replaced the City, had "cats" too, and the 998cc engine option was quietly dropped. By 1994, all new Minis were fuel injected. The cars certainly resembled living antiques on the surface, but

John Cooper and 1960s race ace John Love are all smiles at the launch of the Mini Cooper 35 special edition in 1994.

under the hood at least they were keeping pace with the big emissions clean-up gripping the whole of the global car industry.

It seemed an energy boost had been delivered to the by-then very elderly Mini. Rover added some worthwhile improvements, such as Metro-based front seats and an internal hood release in 1993, together with a very attractive full-width wood dashboard for the Mayfair.

The drive to keep the Mini relevant received an important further boost in 1994. That was the year British Aerospace's proprietorship of Rover came to an end when the British company, with all its bundled-together British motoring heritage, was sold to Germany's BMW. The shock move proved too much for Honda, which abruptly severed its ties with Rover, but anyone with a strong attachment to the Mini—that is, anyone who loved cars that were fun to drive and full of character—had cause to celebrate. BMW was clear from the offset: it felt the Mini was one of the overlooked gems within Rover and almost immediately started to plan for an all-new car. For frustrated Rover engineers eager for such a move and bubbling over with ideas, this was a healthy engine roar to their ears. And not only that, but quite improbably, Bernd Pischetsrieder, BMW's chairman and an avowed car enthusiast, was a distant cousin of Alec Issigonis.

Not long after BMW's arrival at Longbridge, the Rover 100 reached the end of its road, in 1997. Simply the final, rebranded version of the largely uncharismatic Metro, the 100 had now been outrun by the Mini, the car it was meant to supplant seventeen years earlier. For there was absolutely no question of the thirty-five-year-old Mini itself being finished off—keeping it going would now be a vital part of building up the Mini brand ahead of the arrival of the totally new model. Rover's UK design and marketing teams were challenged by their new Munich bosses to devise as many interesting ideas as they could to sprinkle stardust on the Mini. The original investment in the car had been paid off years earlier; the impetus here wasn't so much to wring a few more pennies of value from it but to adopt the Harley-Davidson approach to stoking desire. That is, there's no rational reason to shell out much more for an old-fashioned Harley motorbike than you would for a modern Honda, yet the image of the easy-riding machines still makes consumers do just that.

The first results of looking again at the classic Mini came quickly, being revealed in October 1996. The changes added up to one of the most radical, and worthwhile, Mini revamps for years. Keeping the car just ahead of the ever-tightening new car-use regulations saw its radiator repositioned to the front of its compact engine bay—something first considered in the late 1960s—to help it meet drive-by noise standards for new cars. Meanwhile, the fuel injection was upgraded to a multipoint system for cleaner emissions. Producing 63 brake horsepower, the same engine was now fitted to both Cooper and 1.3i.

On top of this came an unexpected focus on occupant safety, with a redesigned steering column now housing a driver's airbag and side impact beams fitted inside the doors. A huge amount of sound deadening was added, everywhere from under

[top] The 1997 Sports Pack looked fantastic, although those superwide wheels didn't add much to the handling . . . or ease of parking.

[above] The handsome and restrained cockpit of the 1998 Cooper LE, with a '60s Cooper race-team car in the background. There was now an airbag in the steering column.

ROOFLESS CHARACTERS

The first ever convertible Mini was built in 1962 by two friends working for an engineering company in Croydon. Jeff Smith and David McMullen sliced the roof off the car, then added plenty of strengthening under the sills and the back seat to keep the car from flexing and beefed up the door locks to prevent vibrations. The reaction to it was so positive that they left their jobs and started Crayford Auto Development in Westerham, Kent, and began selling replicas through dealer chains from 1963. They were soon building two a week, and the cars were much improved when the two decided to leave the original side door and window frames in situ to make the car even stiffer and compensate for the strength lost by cutting the top off.

The company became quite famous in 1966 after food firm Heinz commissioned fifty-seven Crayford-modified Wolseley Hornet convertibles for a major consumer competition. As Smith and McMullen were trained engineers, their work could be trusted. But that wasn't always the case for other chop-top Minis. In the 1980s, conversions proliferated in both the United Kingdom and mainland Europe, and some of them were highly suspect when it came to integrity. One company that did make a very good fist of letting sunshine into a Mini was German dealer LAMM Autohaus. In 1991, Rover was so impressed with its work that it gave its blessing to a run of seventy-five cabriolet convertibles. Not since the Moke of 1964 had there been such a radical new body style in the catalogued Mini range, even if the stacked-up hood gave a somewhat pram-like appearance when folded. When the cars had all sold, Rover decided to buy the project from LAMM in Baden. The German soft-top experts of Karmann were called upon to comb through the design and prepare it for much wider general sale in 1993, and it went for an eye-watering £12,000. The body was more thoroughly reinforced (adding 154 pounds to the weight of a standard Mini Cooper), and a professionally designed and manufactured hood was supplied by renowned British convertible specialist Tickford.

Costly through that was, Rover still managed to sell 1,081 examples up to 1996 . . . and deterred a lot of people from buying a soft-top Mini built by a man in a shed.

[top] Crayford founder David McMullan with one of the several hundred Minis that his firm opened up for fresh-air fans.

[above] Rover's Mini Cabriolet, launched in 1993, was a factory-built first; the high-quality hood had to stack up at the back in a somewhat pram-like fashion when folded. £12,000 was a lot to ask, but then plenty of work went into properly reinforcing the Mini Cabriolet so it retained integrity . . . unlike some of the many unofficial conversions.

[top] The overstyling would have had Issigonis fuming, but the striking dashboard of the Mini Cooper Sport of 2000 looked great to everyone else.

[above] Wood interior pack, wooden gear knob, and a full stone leather retrim—a great way to spec your safer, cleaner post-1997 Mini, if you could afford it.

the roof to around the fuel tank, to make the Mini less of an assault on the senses over long distances, and every model now came with a full-width wood dash.

The familiar Mini profile, naturally, remained untrammeled. What was different was a vast catalogue of parts and accessories now on offer so you could custom-create your own unique car. No longer would there be a new limited-edition model every few months (aside from the Paul Smith and fortieth-anniversary cars) because almost every Mini would now be highly individual. That could stretch from a dramatic full-roof checkered decal to the Sports Pack with its bank of spotlights and super-wide wheels, tires, and spats (some of the wings had to be cut away to fit the latter)—a little overwhelming for the hard-pressed suspension and the driver trying to park, but nonetheless glorious looking.

The finishing touch was the Mini's first proper marque badge, a hood emblem with flowing wings on either side of the Mini legend in the manner of British sports cars from Aston Martin to Morgan.

The car now started at a costly £8,995 before any of the fancy new extras were specified. It was no longer positioned as a cheap starter car but as a fully priced style icon on which a buyer could easily lavish fifteen grand. Consequently, and maybe entirely as BMW anticipated, sales fell sharply, settling at an annual output of about ten thousand cars, half of them bound for Japan. In 2000, the year that included the original Mini's final ten months on sale, just over seven thousand were sold.

The Mini had ridden the ups and downs of the British motor industry all its very long life, and there would be one more pothole in the road for it to struggle over. In March 2000, it was sensationally revealed that BMW was in talks to dispose of the MG and Rover marques and the huge Longbridge factory after having found the mainstream car business was leeching away its profits and energy. There was uproar

in the British press, and in the end the business was sold not to the venture capitalist Alchemy Partners but to the Phoenix Consortium, a group of Rover managers past and present that had pledged to turn the business around and save both the marques and the many jobs involved. Land Rover was sold off to Ford, but BMW retained Mini. In a dramatic swap, the production lines for the upcoming new Mini were uprooted and moved to Cowley, Oxford, while the Rover 75 plant located there was shifted over to Longbridge.

Throughout all of this, the old Mini continued its dwindling life, and all responsibility for making it and serving its existing owners transferred to MG Rover Group. And so it was that Kevin Howe, the managing director of that newly formed company, was on hand as the very final original Mini was completed on October 4, 2000.

It truly was extraordinary that the car had driven into the twenty-first century. In 1959, its launch year, about twenty thousand examples of this revolutionary small car had reached uncertain yet excited customers. The first million was reached in 1965,

The wide-wheeled, gold-painted Knightsbridge final edition outside a similarly theatrical junk shop on London's Portobello Road (about 2 miles away from Knightsbridge).

[top] The final-edition Cooper S with all the right bits was guaranteed to be a collector's item as soon as you took delivery in spring 2000.

[above] The last of the plain and simple Coopers, seen here in March 2000, was every bit as appealing as you could hope for in verdant green with a white roof.

the year the 70-mile-per-hour speed limit arrived in Britain (that was the top speed of an 848cc Mini, anyway). The second million was celebrated in 1969, as man stepped on to the Moon and the supersonic Concorde airliner took to the sky. The third million came three years later, in 1972, despite new superminis like the Renault 5 setting the pace in the small-car arena. In 1976, the four-millionth Mini was made, and only in that year did it receive a heated rear window and hazard warning lights as standard kit. The five-millionth Mini burst into life in 1986. But there were fourteen more years of life in it yet, and as manufacture ceased its total production tally reached 5,387,062. It is, by a massive margin, the best-selling single British car of all time, and its hordes of ardent fans may have shed a few tears at its demise. But a brand-new chapter of Mini history was just about to open.

CHAPTER

CREATING A NEW MINI

AIRBAG

WITH BMW NOW IN CHARGE, AN ALL-NEW CAR WITH A BRAND-NEW PHILOSOPHY CAME TO LIFE. THERE WAS MINI CHARISMA IN ABUNDANCE, BUT ALSO A THOROUGHLY MODERN APPROACH TO WHAT THE CAR WOULD ACHIEVE AND MEAN.

BMW's purchase of Rover Group in 1994 triggered a merger mania across the German motor industry. Mercedes-Benz soon acquired Chrysler in the United States, and Volkswagen swelled its portfolio with marques as diverse as Škoda, Bentley, and Lamborghini. In 1998, BMW pulled off an even more audacious move when it grabbed Rolls-Royce Motor Cars. For years, cash had been pouring into Germany's coffers as the world couldn't get enough beautifully built premium cars; now it was being spent in a campaign to mop up any available car company with a great name that was showing signs of neglect and mismanagement.

The frenzied round of takeovers, though, tended to come with difficult periods of adjustment—none more so than BMW's imposition of its will on Rover's culture. The nonappearance, despite numerous plans, of a new Mini some thirty-six years after the car was launched was certainly a missed opportunity of epic proportions. Everyone recognized that. Now, that was all set to change.

One of the earliest sketches from 1995, by Frank Stephenson, showing broadly the shape of the new Mini.

The desks and pinboards of Rover designers had been festooned with early thoughts on a totally new Mini even before BMW's arrival. Yet not far away, the old car still rolled off the line; it was highly labor intensive to make, with an engine dating back to 1951 and at the very limits of legislative legality, and a body structure that, from the viewpoint of passive safety in an accident, was woefully behind the times.

When BMW arrived at Rover, it was pleased to find some inspired work already underway. This included real-life stuff, such as an original Mini replumbed with the second-generation Metro's interconnected Hydragas suspension, and Dr. Alex Moulton's enthusiastic involvement. Meanwhile, in the proposal stage were all manner of thoughts, allegedly including a city car with a central driving position and two passenger seats behind—something like a tiny, low-powered McLaren F1.

A thumbnail mood board by Frank Stephenson of the finer points of his Mini.

BMW chief Bernd Pischetsrieder was adamant about harnessing this British creativity. "I want to make it clear that Rover's and Land Rover's design and engineering operations will remain fully functional and largely independent from us here [in Munich]," he was quoted as saying. The new Mini was assigned a project code, R59, and it was all systems go for the ideas lab, with Pischetsrieder keen to involve key surviving people from the original Mini's formative years, including Jack Daniels, Alex Moulton, and John Cooper (although in the end their input would never really figure in the final car).

Perhaps inevitably, Rover people wanted to create an economy car in the pioneering spirit of ADO15. Mini design director David Saddington worked on a classlessly modern four-seater update of the familiar shape on the old Mini's 10-foot length while colleague Oliver Le Grice conceived a more radical one-box design with a rear engine mounted under the floor and using subframes. Meanwhile, to hedge their bets, they also called in Roy Axe, a consultant with many years of experience inside Rover, to shape a more conventional supermini in the contemporary European idiom.

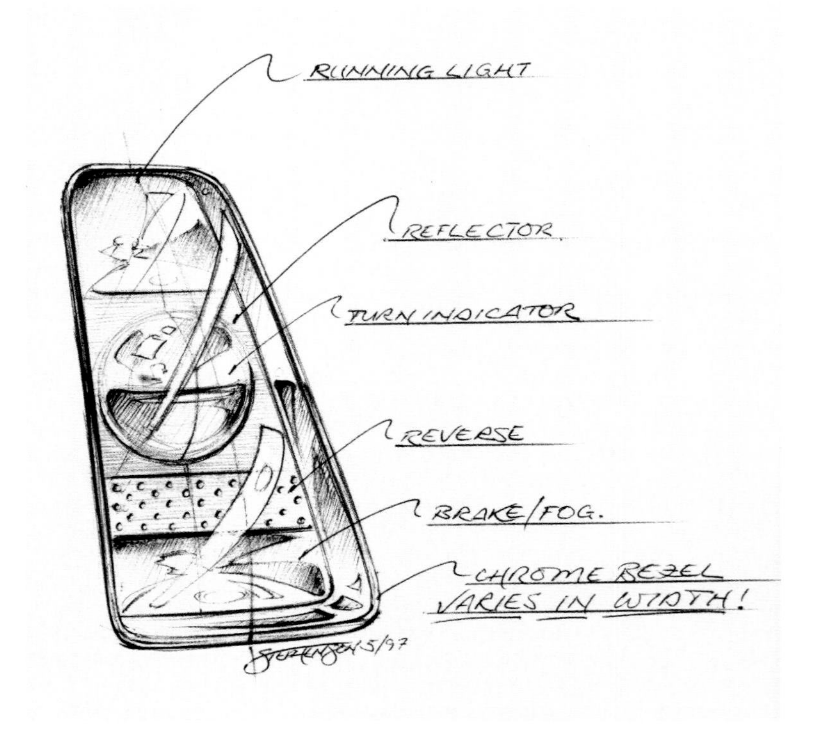

[top] Stephenson's proposed Cooper rear end had, in 1995, the number plate below the bumper line.

[above] Frank Stephenson's winning Mini design given some sparkle, and a tiny door mirror.

[right] Intricate design notes on rear light cluster.

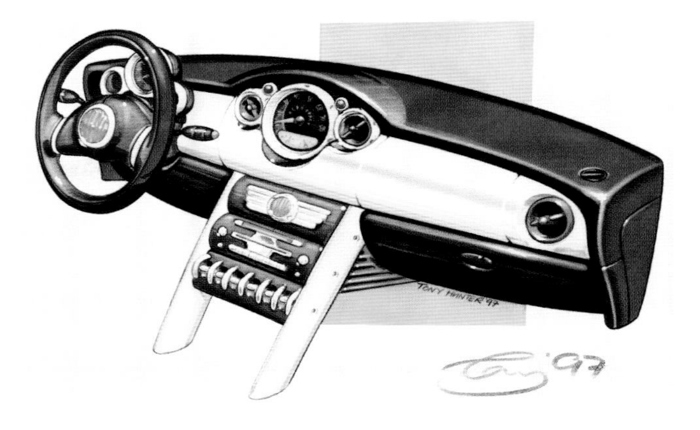

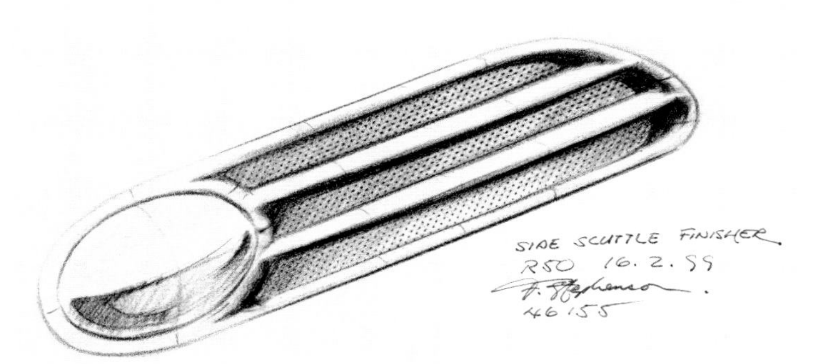

[top left] Interior design concept artwork from 1997 showing the overriding theme of circles and ovals.

[top right] A 1997 rendering by Tony Hunter of the dashboard and center console.

[above] Side finisher added character and visual interest to the Mini's flanks.

A date was set, October 15, 1995, for a decisive design conference to be held at the Heritage Motor Center in Gaydon, Warwickshire, right next door to Rover's test-track facilities and organized by Saddington. It would be an international meeting of minds, and not just because BMW senior management was in attendance; despite assurances that the Brits would be designing the new Mini, BMW had also asked its own design team in Munich and its Californian Dreamworks satellite to examine the brief too. Unencumbered by almost half a century of Mini traditions, these groups viewed it from a true outsider's perspective. The instinctive reaction of design chief Chris Bangle and his team was to imagine a new Mini Cooper, a compact sporting car rather than a 10-foot-long city car. What's more, a Rover 75–style suspension set-up of MacPherson struts at the front and Z-axle at the back was favored, especially by BMW product guru Wolfgang Reitzle.

The Gaydon summit would prove decisive. With three designs from Rover and several more from BMW on hand, the debate was intense. The Le Grice design was soon out of the running, deemed brilliant but too futuristic, and eventually it was down to David Saddington's proposal and a coupe-like car modeled by American BMW designer Frank Stephenson. Rover staff may well have felt betrayed, but BMW, as a supremely accomplished car company, was determined to produce a car that would be both popular and profitable. A conventional small car with sporty and retro overtones would provide that, it felt.

So it was decided that Saddington would manage the design project, with Stephenson working on his exterior, while Rover designers Tony Hunter and Wyn Thomas were in charge of the interior. Saddington oversaw the packaging to turn it into a full four-seater, while BMW in Germany would get on with the engineering tasks. It soon became clear, though, that all the activities of the project needed to be together, and Rover took back the engineering work in May 1996. The new Mini was now called R50 and would come to life as a British car using Rover's K-series engine.

Or, that's what engineers in Britain assumed until BMW made the sudden announcement that a brand-new engine would be provided from a new joint venture with Chrysler and sourced from a factory in Brazil. This was a serious dent to British

Stephenson hard at work on the Mini full-size clay design model.

pride, but there were apparently already problems squeezing the K-series under the low hood line of Stephenson's design, and although the K-series was a very responsive engine, it was known for reliability issues and warranty claims that, perhaps, BMW wanted to avoid in its new small car. Nevertheless, Rover's existing R65 five-speed gearbox was adapted to mesh with the new engine, and the team performed miracles to get the steering and front suspension all neatly packaged under the car's tiny nose. Nonetheless, communications with Munich didn't always run smoothly, and reportedly at one time a firm of industrial arbitration specialists was summoned to help foster understanding. This interjection was viewed with typical British cynicism, and the suspicions that BMW was actually exercising ultimate control over the Mini design process were vindicated after both Pischetsrieder and Reitzle left the company. Pischetsrieder's departure, in particular, was a tough blow; he was very pro-British, having once said, "The Mini is the only lovable small car. The others are just like bars of soap." Most of the final testing of the near-production-ready car was taken away from the British team in 1999 and relocated to Germany, where much refinement was done at the Nürburgring circuit so familiar to BMW engineers.

The development of the new Mini probably was more fraught than for most all-new cars; after all, how could it not be when the creators were replacing an icon and when history was already littered with many false starts on just such an undertaking?

[top] Mini design team leader Gert Hildebrand discussing paintwork with Marisol Manso-Cortina, in charge of color and trim on the project.

[above] David Saddington worked on the packaging of the new car, as well as on British design proposals that competed in the initial design contest.

For Rover, though, there was the additional stress of its seemingly always-unsettled ownership. In May 2000, BMW disposed of the mainstream Rover car business while carefully extracting the new Mini project from it. This meant abandoning the new, purpose-built Longbridge assembly hall and relocating the new Mini production line to Cowley, Oxford. This was something of a homecoming, as more than six hundred thousand of the original Morris Minis had been made there between 1959 and 1968.

BMW's exit from Rover dominated the news agenda in the spring and summer of 2000, but the new, nearly customer-ready Mini Cooper—called, officially, the MINI Cooper—somehow managed to transcend the corporate turmoil to make a decisive and positive public debut at the Paris motor show in October that year. Designer Frank Stephenson summed it up when he stated baldly, "It has the genes and many of the characteristics of its predecessor, but is larger, more powerful, more muscular and more exciting."

More exciting? That was a little harsh on all those great characters who had contributed to the original car that Stephenson was then able to so adroitly pastiche. Both Mini purists and industry sages decried the way the new car abandoned so much of Issigonis's thinking, especially in its weight, its size, and the wasteful packaging approach that made the four-seater accommodation very tight indeed.

With the original car about to be laid to rest, and the new one already familiar to most enthusiasts, a new era was dawning. The MINI was the first premium supermini.

But what of the car that greeted buyers when it went on sale in the United Kingdom in July 2001? The first impression was of a vaguely familiar profile that had been expanded in all directions. The MINI adhered faithfully to the wheel-at-each-corner

stance, but it was longer, wider, and chunkier, with the emphasis on its enormous (relative to the older car) wheels. The car was almost 2 feet longer than its predecessor. As the overhangs were minimal, the extra space went into the cockpit and to state-of-the-art safety systems and crashworthiness. The gently rising shoulder line of the car was established at its round headlights, while the broad, hexagonal grille shape picked up on the form familiar from all original Minis since 1967. The upright rear light clusters and chrome surrounds immediately evoked the 1959 original. On the other hand, the interior accommodation was almost like a 2+2 GT, with the emphasis on the two front seats at the expense of the relatively tight rear ones and on a gigantic circular "speedo" dominating the middle of the dashboard and housing all the car's dials and gauges. There was, of course, a hatchback and folding rear seats—something glaringly missing from the original Mini even in the mid-1970s!

To pack everything under the very slight front end, a clamshell hood took the top sections of the front wings, and the headlights, with it when it swung upwards. And below it was the four-cylinder, all-alloy, 16-valve Tritec engine at 1,598cc that the car would share with the Dodge Neon. It was a retrograde step compared to the K-series in that only the cylinder head was aluminum, the block being old-fashioned cast iron. It came in two states of tune, with 90 brake horsepower for the base-model MINI One (£10,300) and 115 for the mainstream Cooper version (£11,600) around which the entire new MINI venture had been conceived.

The old car had only ever offered a four-speed manual gearbox or a conventional automatic transmission, but now there was either a five-speed manual or electronic continuously variable automatic; the latter optionally featured a semiautomatic button shift control on the steering wheel with six firm ratios, called Steptronic, for anyone averse to the decidedly unsporty note of a CVT transmission. Of rubber suspension systems there was no trace, with MacPherson struts at the front

[top left] The 1.6-liter Tritec engine for the new Mini was co-developed with Chrysler and would be sourced from Brazil.

[top right] Front and rear suspension systems for the new car drew heavily on BMW practice.

[above] The state-of-the-art safety structure of the all-new Mini was a huge leap over the original car's.

[opposite top] Chilly headquarters for the MINI's cold-weather testing team.

[opposite bottom] Hot-weather testing to ensure the MINI could keep its head wherever it was driven.

and a multilink rear axle. There were disc brakes front and back, with electronically controlled stability and brake-force application programs as standard within the antilock braking package, and a novel drive-by-wire throttle. Electronic traction and stability control were optional. It was the first production car with a puncture-spotting tire-defect indicator as part of a comprehensive safety package packed with front and side airbags and optional run-flat tires. One of the visual signatures of the original Mini Cooper had been a white-painted roof to contrast with the overall body color, and this was back on the MINI Cooper and integral to its look, with the darkened windows wrapping around the car like a continuous band of glazing. The Cooper also featured a hood ventilation slot and twin exhausts. Distinctive it most certainly was.

THE LONG, LONG TEASER

By the mid-1990s, the "concept car" had become a staple of the annual circus of new car launches and motor shows. There are two types: the wild flight of styling and technology fancy that demonstrates the manufacturer is bursting with creative energy for what might be to come in the medium to long term; and the close-to-reality preview of a new production model soon to be launched in toned-down form. This latter concept car acts as a cushion to soften up the shock of the new—to get the buyers prepared for what's to come in the short term. Both, really, are marketing tools, and they're not prototypes because concept cars are generally not runners.

After decades during which British Leyland had been necessarily guarded about its future plans, from 1994 onwards under BMW's ownership it wanted to signal that, while the classic Mini was being treated to a late-life fillip, there was a brand-new car on its way . . . and that buyers might want to defer buying any other new car until it arrived.

BMW, however, chose an unusual route to signaling its intentions with Mini concept cars. It started to use design rejects to build interest.

First, at the Monte Carlo Rally in early 1997, it revealed the ACV30, a tiny two-seater coupe that had begun life as one of BMW's own Californian-created Mini proposals from the Gaydon summit two years previously. The timing was sweet in that it came thirty years after the Mini's final victory on the event. It apparently used MGF running gear and had some retro overtones of the classic Austin-Healey Sprite Mark I "frogeye," with its headlights set cheekily on top of the hood along with bulging wings and a tight cabin. It was proof positive that BMW saw the new Mini through sporty, Cooper-tinted glasses from the start. The interior, though, was prophetic, with its painted metal and central circular instrument display.

Just a few weeks later, Oliver Le Grice's pair of rear-engined Mini proposals—other nonstarters from 1995—appeared in public for the first time at the Geneva Motor Show. The three-door car was now named Spiritual and its longer five-door companion the Spiritual Too. If the reaction was lackluster, then it only confirmed to BMW that its suspicions were correct about what people wanted an all-new Mini to be like.

Finally, at the Frankfurt Motor Show in autumn 1997, a mockup of the near-final MINI Cooper was unveiled. Surprisingly, this one did actually work, it is said, as underneath the profile that would soon become familiar globally there were the underpinnings of a humble Fiat Punto. It was, however, to be a very long wait—almost four years—before customers could get their hands on one of the first production cars . . . driven, BMW must have hoped, half crazy by the anticipation.

[top] Issigonis-like sketches of Spiritual, showing the freehand packaging benefits of a rear-mounted engine.

[above left] The ACV 30 concept car was unveiled in March 1997 to mark thirty years since the Mini Cooper won the third of its Monte Carlo Rallies. Bulging wheel arches and roof profile suggested an organic look that, in the end, wasn't pursued for production.

[above right] Spiritual (foreground) and Spiritual Too represented the British outlook on a new Mini as the advanced economy car reinvented. Seems a shame that the Spiritual didn't progress beyond this show car—BMW said, in consolation, that it was ten years ahead of its time.

[opposite] The almost-final shape of the new MINI Cooper was revealed in 1997—a culture shock to longtime fans of the British car. BMW came to regard any new Mini as actually a new Mini Cooper, and made the rally-inspired livery integral from the start.

CHAPTER

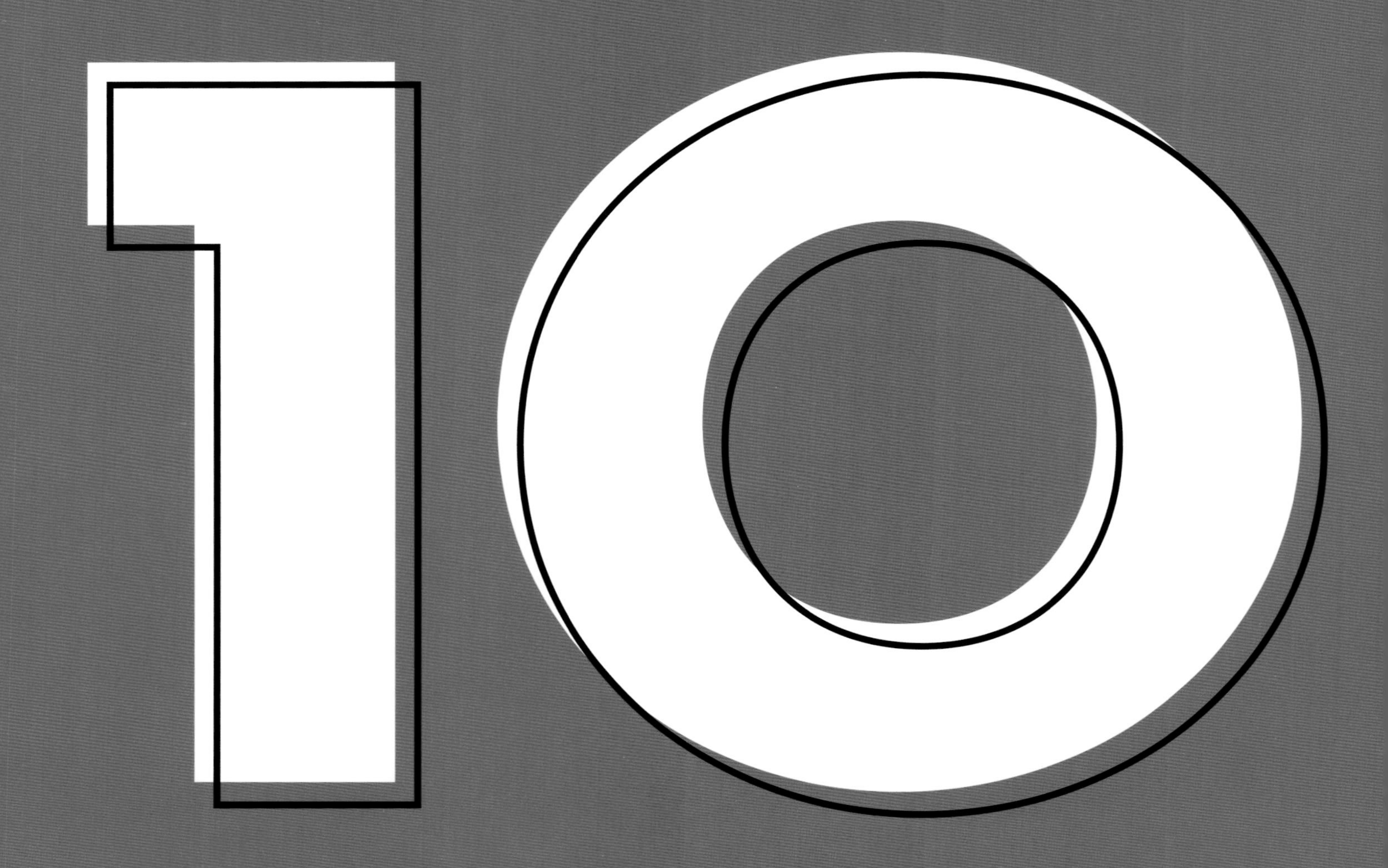

MINI, A CLASSIC REINVENTED ENTIRELY

854

WITH THE COOPER AT THE FOREFRONT FROM THE START, THE NEW MINI SET THE SMALL-CAR WORLD ALIGHT ONCE MORE. IT WAS AN INTERESTING AND DESIRABLE SMALL CAR AND A HUGE SHOT IN THE ARM FOR BRITISH MANUFACTURING.

The floating roof idea, with pillars and glazing forming a dark band around the top part of the car, was adapted from the original Range Rover for the new MINI.

That this should be the most eagerly anticipated new car of 2001 was hardly a surprise. The motoring media was anxious to get its hands on the MINI to relay what it hoped would be good news to potential customers champing at the bit.

It was a testimony to the combined engineering skills of both the German and the British development teams that the MINI was discovered to be such a driver's delight. Every critic who got behind the wheel was fulsome in his or her praise for the car's wonderful handling and pin-sharp, expertly weighted steering. Venerable weekly magazine *Autocar* succinctly praised "its ability to involve you in the action and even let you alter your cornering line using both throttle and steering. Yes, there's a hint of lift-off oversteer in extremis, but mostly there's so much grip that the only slip you're likely to encounter will be mild and at the front axle if you push really hard through a tight corner." The brakes came in for wide praise too. Some may have grumbled at the car's bulk and over-obvious trendiness, but the old Mini's tendency toward terminal oversteer on tight bends taken too fast had been confined to history as the new one delivered enormous fun with the limits of its road adhesion extended to far beyond where even most over-enthusiastic drivers would dare to probe. "This is a car you aim through corners confident that you're going to clip a blade of grass," said *Autocar*.

The excellence of the chassis was not in doubt—which made the mediocre performance of the engine all the more obvious. With 1.16 brake horsepower to haul along a car weighing 1,125 kilograms, even the Cooper was found to be a lackluster goer by everyone who tried it, with a mundane 0-to-60 sprint time of 9.6 seconds and decidedly tardy acceleration through the gears in midrange urge. Winding the car up to 100 miles per hour took more than 28 seconds.

[left] The soft-hexagon grille of the new car carried on the outline shape first seen on the Mini Mark II range in 1967.

[bottom left] A key feature inside was the centrally positioned instrument display, circular just as Issigonis envisioned, although with a modern console below.

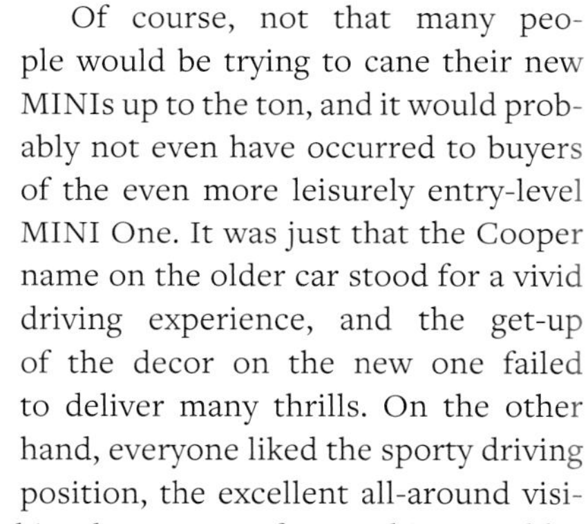

Of course, not that many people would be trying to cane their new MINIs up to the ton, and it would probably not even have occurred to buyers of the even more leisurely entry-level MINI One. It was just that the Cooper name on the older car stood for a vivid driving experience, and the get-up of the decor on the new one failed to deliver many thrills. On the other hand, everyone liked the sporty driving position, the excellent all-around visibility, and the ambience and finish of the cabin; there were a few teething troubles and a few minor recalls, but the BMW dealers selling the new MINI acted swiftly to help the factory iron things out so that customers didn't ever experience the dismay and exasperation they might have felt during the darkest days of British Leyland's travails in the 1970s. The novel Salt, Pepper, and Chili packs proved hugely popular.

In January 2002, BMW swung into action to fix the criticism of the MINI Cooper's underwhelming performance. The range was boosted—literally—by the addition of the high-performance Cooper S. Its lightning acceleration of 0 to 62 miles per hour in 7.5 seconds, with a 135-mile-per-hour top speed, was down to a supercharged version of the Tritec engine that offered 163 brake horsepower. A six-speed manual transmission was offered on the Cooper S, together with a taut sports suspension package. The Eaton supercharger fitment meant the battery had to be housed in the trunk (shades of the old Mini there again), which meant sacrificing the spare wheel, and therefore run-flat tires were fitted as standard.

As ever with new MINI developments, there was a long gap between announcement and availability, and the Cooper S entered showrooms in June 2002 with a £14,500 price tag. In fact, the supercharged car had been part of the plan from around 1998, but its arrival was rather timely.

The Cooper and Cooper S, of course, were regular production models, which flowed off the regular MINI production line at Cowley. BMW and John Cooper had agreed on a licensing deal for authorized use of the Cooper name, and although Cooper sadly passed away in 2000, John's son Michael maintained the connection.

MAKING THE MINI AT COWLEY

Back in 1996–1997, BMW had spent £280 million modernizing the former Morris factory at Cowley, Oxford—first opened in 1913—in preparation to build the new Rover 75. The body shop and assembly lines that had once played host to the Morris Minor, Morris Marina, and Rover 800 were fully revamped, while a brand-new paint shop was at that time the second-biggest construction project underway in the United Kingdom after the ill-fated Millennium Dome.

Then came the difficult task of moving the Rover out and moving the MINI in, which came with £230 million more investment in 229 robots to build the cars' bodies "in white" (sealed, primer-coated, and ready for final painting) and a laser-guided measuring system to give body assembly accuracy to the nearest 0.05 millimeter. This may have all been run-of-the-mill stuff for German cars, but for the British motor industry it meant setting painstakingly stringent new standards; exactly 602,817 examples of the original Mini had been produced here up to 1968, but that was at a time when no one really knew what a millimeter was . . .

From the time the decision was taken to base the MINI at Cowley to the first series production cars coming off the line in June 2001 was thirteen hectic months, and the job required help from workers at BMW's Regensburg site in Germany. Once the go button was pressed, with 2,500 hands on deck, Cowley was rechristened Plant Oxford (even a boring old car factory could be rebranded to sound efficient and brainy). The paint shop had been replanned to allow for the two-tone paintwork called for on MINI Coopers, and an innovative Core Production Integrating Management System ensured that near-one-off specifications would be built into each individual car as it moved through the manufacturing process. Flexibility like this ensured all the myriad options the MINI customer had selected in the showroom a few days or weeks earlier could be achieved seamlessly and without scrappy bits of paper stuck to each windscreen.

Plant Oxford showing the stage of manufacture where engines and transmissions were installed into completed body units.

[above] The clamshell-style hood took the wing tops and headlights with it when raised.

[right top] A full-length union jack as a roof motif was a popular option, especially for Coopers.

[right bottom] Rear accommodation was tight, as in any close-coupled sporty car, but perfectly adequate for two children..

He established a new company called John Cooper Works with the aim of coaxing even more performance out of the standard Cooper S. Launched in mid-2002, the first JCW tuning kit could only be fitted to Cooper S cars already registered for road use; with a gas-flowed and ported cylinder head, a faster-spinning supercharger, and an uprated exhaust system, plus a remapped ECU to cope with the changes. The car could now generate 200 brake horsepower and sprint to 62 miles per hour in 7.4 seconds. The brash John Cooper Works livery broadcast all this, of course, and every car came with an individual number and certificate. *Autocar* magazine declared it "a gloriously frantic affair, goading you into driving in classic all-out Mini style," and loved the loud exhaust noise and the engine's rev-happy character.

Many years before the "Dieselgate" scandal blew apart any environmental credentials that might have attended it, diesel was an essential part of pretty much every new model range. BMW clearly felt the MINI could be no exception, and so in June 2003 it made the MINI One D the first ever diesel variant of Britain's most [right] The three-car MINI range in 2002 (left to right): One, Cooper, and Cooper S.

[below top left] Thanks to its standard supercharger, the 2002 Cooper S added meaty performance to the MINI.

[below top right] The MINI One D mixed the base-level spec with the miserly fuel consumption of a diesel engine.

[below bottom left] This experimental hydrogen-powered MINI, revealed in 2001, was part of BMW's alternative fuels investigation program.

[below bottom right] John Cooper Works tuning kits were launched in mid-2002, in happy partnership with John Cooper Garages in West Sussex.

famous economy car. The 1.4-liter turbo-diesel 75-brake-horsepower engine was supplied by Toyota, fresh from its recent use in the French-built Yaris supermini. The common-rail technology inside its compact, all-aluminum package produced a MINI that could go a long way between return visits to the fuel pump, with thrifty consumption of 58.8 miles per gallon.

This eye-catching MINI stretch limousine, a one-off with a built-in jacuzzi, made waves in 2004.

In the spring, appropriately, of 2004, the next MINI chapter opened with the launch of the Convertible. If Issigonis's ghost had been clanking his chains in frustration at the way his original design had been reinterpreted, he might have found some solace here: unusually for its size, the car came with a power-operated roof that stowed, when folded, high up behind the rear seats, which allowed trunk space below it to be accessed through a drop-down lid very similar to the one found on the first Minis in 1959. Just as for Issigonis's original, the simple hinges were external, only rather than being painted to hide the money-saving fix, they could optionally be chrome-plated to emphasize them as a design feature—clever, that.

The hood itself was a high-caliber fabric affair, multilayered and insulated, and incorporated a glass rear window with heating element. The small rear passenger side windows retracted automatically when the hood was lowered. In its overall design the Convertible was, really, more of a cabriolet in the tradition of the old Volkswagen Beetle, with everyday practicality more important than supersleek lines with the hood down. However, the hood section could also be slid back electrically for a coupe de ville feel—i.e., a deep, full-width sunroof—on sunny but crisp winter days. The Convertible body style could be had with One, Cooper, or Cooper S specifications. In a probable bid to endear the car to the wealthy female demographic, the launch was accompanied by a Convertible covered in a "dress" consisting of thirty thousand glass mosaic gems by Italian fashion brand Bisazza. This was just one of many media-attractive events BMW's MINI marketers organized to push the car under the noses of customers well outside car-enthusiast circles.

Carrying on the personalization program begun with the original car in its late-1990s sunset days, BMW created a vast array of accessories and options so almost every car could leave the production line custom-finished for the buyer. The graphics,

O- [right] Four-seater open-top fun in the MINI Cooper convertible, unveiled in 2004. A power-operated fabric roof was standard on the convertible (this one is a Cooper).

O- [right middle] This John Cooper Works GP Kit model was a radical 2006 limited edition. Its unique specification was constructed by Bertone in Italy.

♀ [right bottom] Axing the rear seat was a novel weight-saver for the John Cooper Works GP Kit model.

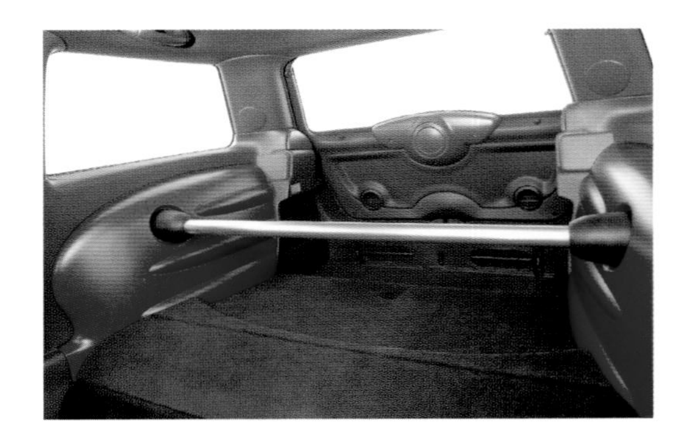

wheels, trim, and interior extras were generally cosmetic, while the John Cooper Works kits took care of added performance. In 2005, the JCW power-boosting kit for the Cooper S became a factory-fitted option, with power inched upwards to 210 brake horsepower, while for those content with the standard car but wanting just the hotted-up aura, the JCW Sound Kit now gave a more growly exhaust note and 3 more brake horsepower just with a revised exhaust. In 2006 came a stripped-back limited edition entitled the MINI Cooper S John Cooper Works GP Kit; to save weight, it had no rear seats, rear wash/wipe, or air con, minimal sound-deadening material, and a carbon-fiber rear spoiler. Each one of the 2,000 made was hand-finished by Bertone in Italy. With 459 of them finding British owners, it's already a collectible car.

The misgivings of diehard Issigonis fans might have caused a stir in the early 2000s, especially on the unfolding forum of the Internet (then still quite a novel place to stand on your virtual soapbox). Whatever they thought, the new MINI was a tremendous success. The five-hundred-thousandth new MINI—a silver Cooper S—was built in August 2004, thirty-seven months after the first had hit showrooms. This sales rate was actually fractionally ahead of the original Mini's. Records reveal that 509,572 had been sold by the end of 1962—three years and four months since it had first become available. The Cooper proved the biggest-selling variant thus far, with 250,000 shifted, and MINIs overall trounced premium rivals from BMW's German archenemies, such as the technically advanced Audi A2 and the innovative Mercedes-Benz A-class.

[top left] The MINI One Seven of 2005, a special edition plundering the Mini's earliest days for a catchy name.

[top right] The MINI Concept show car of 2005 previewed the new estate car still to come.

[above] How to get the fresh air to rush past faster—drive a MINI Cooper S convertible, seen here in 2004.

The starkest contrast between the former and current Mini/MINI was in exports. Some 284,000 of the half-million-plus-a-few 1959–1962 original Minis were sold abroad, but BMW had already shipped 375,000 of the half-million new ones. This included cars to the United States, a territory where the original Mini had made precious little impact during its brief and inglorious spell on sale in the 1960s (all were Coopers or Cooper Ss because the One wasn't powerful enough to support standard air conditioning). Understandably, then, in 2003, new MINI Coopers appeared in a remake of the 1960s movie classic *The Italian Job*, much of it shot around Los Angeles. Like most cinematic reboots, the Charlize Theron/Mark Wahlberg retelling suffered terribly from comparisons to the original, with a critical panning and contempt from fans of both the first car and the first production. Another comedy movie remake of 2007, *The Heartbreak Kid*, starring Ben Stiller and Malin Akerman and a MINI Convertible, was even more of a turkey. Affection simply cannot be bought by product placement, and in trying to make a car lovable and hip through synthetic means, BMW was sometimes seen to be flailing about in its efforts.

CHAPTER

THE SUBTLETY OF THE SECOND-GENERATION MINI

M PZ 6229
M AG 6707
M VS 7462
M TW 8397
M GW 9152
M ZC 5390
M FL 3131

IN THE AUTUMN OF 2006 THERE CAME AN ALL-NEW MINI. NOT, IT SHOULD BE SAID, THAT IT WAS IMMEDIATELY OBVIOUS FROM THE OUTSIDE . . .

Even MINI owners struggled to tell the new car apart from the outgoing one. And, as the MINI's characterful profile was one of its key assets in the small-car market, that was just what BMW aimed for.

But before discussing the highly significant differences, it's important to salute the car that (re)started it all and record its astonishing achievements. All together, the R50 Hatch, the R52 Convertible, and the R53 Cooper S had seen worldwide sales of 730,321 cars, of which 176,113 had been delivered to UK buyers. More than 80,000 examples had been Convertibles. This gave the British motor industry an enormous fillip and put paid to accusations that BMW had abandoned the country when it decided to get out of the mainstream Rover car business. Indeed, Plant Oxford was running at full capacity almost from the start. BMW planned to sell 100,000 cars in a full year of production, but by 2005 that output had rather impressively doubled.

Examples of the second-generation MINI moving through the rigorous paint-inspection process at Plant Oxford.

The entire MINI manufacturing process was overhauled yet again in 2005 for the Mark II. A further £100 million paid for an enlarged body-assembly department to take annual capacity to 240,000 cars and a second paint shop in which a new Integrated Painting Process now applied rustproofing and primer together as the first layer of paint. The start of Mark II production in autumn 2006 also pressed the button on the so-called MINI Production Triangle.

Plant Oxford, naturally, sat at one extremity. Meanwhile, it was fed by the former Rover Swindon pressings plant at another, about 43 miles away. There, fifty stamping machines, ranging in pressure from 400 to 5,000 tons, turned out 90 percent of the MINI's steel body components and assembled the vast majority of body sections such as doors and lids.

At the third point of the production triangle was the crucial new source of MINI components. At Hams Hall in Birmingham, on the site of a former power station, was a brand-new plant making a brand-new engine for the second-generation cars. This factory had been producing four-cylinder gasoline engines for BMW cars since 2001. Now—after a £30 million investment—it was working even harder to produce MINI

engines, putting an end to imports from Brazil and, at the same time, increasing the MINI's British content (by value) from 40 to 60 percent.

The new, all-aluminum Prince engine replaced the Tritec range with few tears shed for the outgoing unit. Prince was co-developed with Peugeot, with many of its components manufactured in France and then dispatched to Hams Hall for assembly. The Mark II MINI Cooper was equipped with a 1,598cc unit featuring BMW's Valvetronic fully variable valve-lift control for seamless delivery of its 120 brake horsepower. More responsive but thriftier, the new motor was loaded with BMW's latest technology, including auto start/stop to cut the engine at standstill and brake energy regeneration to harvest electricity to start it up again when the clutch was depressed; features included a gear-shift point indicator, a volume-flow-controlled oil pump, and an on-demand coolant pump.

[top] Revealed in November 2006, the MINI Cooper S looked familiar but was virtually an all-new car, with every single panel slightly different. It was also bulkier up front so it could meet new standards for pedestrian impact safety.

[above] The latest MINI was found to be a very safe car, scoring the coveted five-point full marks in the Euro NCAP crash test program.

The MINI Cooper S abandoned supercharging and switched to a twin-scroll turbocharger—plus direct fuel injection—to get its added zest, which was now engine power of 175 brake horsepower at 5,500 rpm and torque of 177 pounds-feet anywhere between 1,600 and 5,000 rpm, with an Overboost function briefly increasing that to 192 pounds-feet when called for. This made it now a 140-mile-per-hour machine that could sprint to 62 miles per hour from a standstill in just 7.1 seconds. Yet it could still attain 30 miles per gallon.

The One received a 1,397cc Prince engine, offering a milder 95 brake horsepower, or even as little as 75 in an economy edition—only 109 miles per hour possible, but the fuel efficiency of 53 miles per gallon was what mattered here.

Completing the engine switchover was a new 1,560cc four-cylinder turbodiesel, the common-rail fuel-injected DV6 motor co-developed by Ford and Peugeot and used in multiple cars ranging from the Citroën Xsara to the Volvo V40; such is the nature of hidden parts-sharing in Europe's interconnected twenty-first-century motor industry. The unit arrived at Plant Oxford freshly minted from Ford's Dagenham factory, with a then-phenomenal 72 miles per gallon on offer, plus a CO_2 emissions rating of a mere 104 grams per kilometer.

Unlike in the old days, the new MINI was never going to be allowed to get long in the tooth, and the all-new power units came in a car subtly but significantly different. Despite its familiar looks it was virtually all new and so followed the industry

norm of the five- or six-year model replacement cycle—one that made, for instance, the Morris Minor's twenty-plus years on sale seem almost quaint.

Every single panel was new. The raised nose still featured a clamshell hood, but now it left the headlights anchored in position in the body structure when it opened, rather than taking them with it. Most dimensions were within a few millimeters of the old model, although it was 7 centimeters longer, while larger wheels helped draw attention away from the generally chunkier nose proportions (a reshaping forced on the MINI so it could continue to meet frontal safety rules).

And, indeed, the car was even safer than before with its six standard airbags, three-point seatbelts throughout, ISOFIX child seat fastenings at the rear, and central safety electronics to manage its restraint system. The state-of-the-art chassis helped it win the coveted five-star rating in Euro NCAP crash tests. Even more pertinent to the driver were electronic power steering and a six-speed manual gearbox in every model, no matter how lowly, and a steering wheel paddle-shift option. The characterful circular central instrument display remained but was enlarged to house a decently sized satellite-navigation display.

What customers taking delivery of their all-new MINIs probably didn't realize was that BMW had toiled ceaselessly to make the car not only cheaper to build, and therefore more profitable, but even more adaptable to customer whims. To create the flexibility to make almost every car individually different, Plant Oxford gained a new system of Mobile Standard Production Cells, or Mobi-Cells, for each car, along with computer tracking that meant any of the typical car's two thousand parts could be altered as late as six days before the under-construction car began snaking along the Plant Oxford production line. With the first million new cars topped in 2006, two years later annual output reached 232,000, and it would be a MINI that, in 2009, became the ten millionth motor car to be built at Cowley since William Morris set up shop there ninety-six years earlier.

The new MINI Clubman made its much-anticipated debut at the Frankfurt Motor Show in September 2007 after having been well previewed two years earlier as the MINI Concept show car. There was no actual link to the 1969–1980 Clubman, British Leyland's brazenly cut-price attempt at updating the Issigonis Mini for the 1970s, bar the name, but this was the first estate car since the original Clubman-fronted HL

[top left] The turbodiesel engine in the 2006 MINI was now shared with Ford and Peugeot, rather than Toyota, and gave cars like this Cooper D fuel economy of 72 miles per gallon.

[top right] In 2007, the MINI estate car, called the Clubman, went on sale, with its single rear side Clubdoor adding to its family-friendly practicality.

A Cooper John Cooper Works GP, a limited-production issue that could show a clean pair of heels to almost any hot hatchback.

was quietly dropped in 1982. In one key aspect, the Clubman—which was 24 centimeters longer, its wheelbase extended by 8 centimeters—really did evoke the old Travellers and Countrymans of the 1960s. Instead of the tailgate of other shooting-brake-type cars, it had twin rear doors, each hinged at the side, that swung open to take the corners of the car, but not the rear lights, with them. To maximize this novelty, BMW decided to baptize the feature as Splitdoor. It had another cringeworthy moniker for the ingenious rear-hinged side passenger door (on the right side of the car when you're sitting inside it): Clubdoor. Oh, dear. Still, at least it made access to the roomier rear bench seat easier, despite being on the wrong side of the car for stopping on the roadside, maybe outside a school, in Britain.

The practical nature of the Clubman (with its 32.6 cubic feet of cargo room) didn't stop it being available with John Cooper Works (JCW) performance upgrades, just like the standard Hatch (they called this a Hardtop in the United States and Australia) and the new-model Convertible that arrived in 2008, with its soft-top roof

now able to be lowered electrohydraulically while on the move at up to 18 miles per hour and a pop-up function to the rear roll bar.

These JCW cars were based on the MINI Challenge Cooper S race series cars, which meant an engine and turbocharger tuned and reinforced to give 211 brake horsepower and, thanks to increased charge pressure, maximum torque of 192 pounds-feet at anything from 1,850 rpm upwards and a momentary boost to 206 pounds-feet when accelerating between 1,950 and 5,500 rpm. Astonishingly responsive to right-foot urges, and all with an appropriately adapted six-speed gearbox and 17-inch alloy wheels. Such Hatches and Clubmans needed just 6.5 and 6.8 seconds, respectively, to hit 62 miles per hour and could do 148 miles per hour; Convertibles were very slightly slower on both counts. Seventeen-inch alloy wheels, all four with correspondingly bigger, more powerful disc brakes, were standard, plus traction control and electronic differential lock control for optimum power and grip in tight bends. Original Mini owners only had their judgment and bravery to rely on . . .

In January 2010, the MINI Countryman arrived and caused instant consternation among marque aficionados even though a concept version (called the Crossover) had been paraded around in 2008. On the upside, here was a fatter, taller car with its five seats and five doors making it the first MINI that could properly be called a family car. The availability of all-wheel drive on Cooper and Cooper S variants also elevated it to the emerging crossover status, with the grip and stature to make it

This view of the MINI convertible shows the integral rear rollbar behind the rear seats and the tight packaging for the folded hood. A delightful feature of the MINI convertible was its de ville roof that could be electrically rolled back above the driver and front-seat passenger.

AT LONG LAST, AN OFFICIAL MINI SPORTS CAR

In 2011, BMW boldly went ahead and did what no one at BMC, British Leyland, or Rover Group had ever had the guts to do—put a Mini-based sports car on sale.

There was nothing else on the road quite like the MINI John Cooper Works Coupé; it was a strict two-seater, with tight headroom. The rear air dam automatically rose at speeds of over 50 miles per hour.

R59 was the codename for the Roadster, but the two-seater fun car was only on sale for four years, perhaps limited by its impracticality.

The new MINI Roadster and Coupé were strict two-seaters. The Coupé came first, derived from the Convertible but with a tight little hardtop roof over the two seats and luggage capacity increased to 9.9 cubic feet. For the complete low-slung look, the windscreen was raked back dramatically by 13 degrees over the Convertible's, making it very snug inside. The car had a rear air dam that deployed automatically at speeds of over 50 miles per hour to ensure downforce pinned the slight hindquarters of the Coupé firmly to the tarmac, for the JCW version (Cooper, Cooper S, and Cooper SD engines were also offered) was a lightweight and rapid car, sprinting to 62 miles per hour in just 6.4 seconds and roaring on to a top lick of 149 miles per hour.

No sooner had MINI lovers gotten accustomed to the radical Coupé (R58) than the Roadster (R59) sped up alongside it a year later, with a tight fabric roof that could be either manually or electrically operated, depending on the market.

As neither of these cars was very practical, certainly by comparison to the MINI Hatch, their appeal was limited. Possibly BMW came to realize that the clamor for a MINI sports car had been ignored in the past for this very reason, for when these two-seaters were discontinued in 2015 they were not replaced. Which was indeed a shame.

a compact alternative to a Land Rover Freelander in the desolate wastes of tree-lined suburbia. The regular MINI's engine range was carried over and the family resemblance maintained by using huge wheels and black wheel arches to mask the extra bulk.

The downside? Well, early rumors that it was nothing but a thinly disguised BMW X1 were soon dispelled. They were entirely different cars; the MINI engines were mounted transversely, while the BMW's were inline. Nor was the Countryman built in Leipzig like the X1, though it *was* assembled in Austria on the premises of contract vehicle builder Magna Steyr and so became the first Mini/MINI to be sourced entirely outside the United Kingdom from the start.

Nonetheless, in the spirit of a company renowned for the thought it puts into everything it does, BMW had a plan to make the Countryman itself an icon. In July 2010 it announced the chubby 4×4 saloon would be going rallying in the spirit of the original Mini's glory days of the 1960s. The full assault, orchestrated by those former Subaru miracle workers at Britain's Prodrive, swung into action in the 2011 World Rally Championship with the Countryman JCW WRC, but the relationship between BMW and Prodrive proved a fraught one. A budget overrun couldn't be offset by extra sponsorship. The team contested every event in 2012—there were even a few minor wins, in Qatar and Lurgen Park—and that probably homologated the car as a proper contender, but after that the factory support ended and the WRC was offered

[top] The lurid red highlights mark this interior out as a John Cooper Works edition of the MINI Countryman, with the rev counter now relocated to the top of the steering column housing.

[above] The Countryman, introduced in 2010, was a larger, five-door MINI, offered with four-wheel drive and here seen in Cooper form.

The first MINI rallying adventure was quite short-lived for the WRC car, seen here in spectacular airborne action.

to privateer teams to campaign. Sadly, there weren't any major takers, and the project fizzled out. It was one past glory, at least, that BMW had found it hard to revisit, and the new MINI's rally career was over before it had really begun.

Perhaps BMW was most successful when it shook up its MINI with nontraditional elements, such as the new 2-liter turbodiesel—the German manufacturer's own engine this time, first seen in the BMW 1-Series—fitted in the Cooper SD of 2011, with 145 brake horsepower on tap. And then there was the Clubvan of 2013–2015, a business-oriented commercial version of the Clubman with paneled-in rear windows. It did modestly well in Europe but hit problems in the United States, where tariffs slapped on commercial vehicles—the so-called retaliatory chicken tax for Europe barring poultry in the 1960s—meant the Clubvan found just fifty buyers before it left the US lineup.

As the years rolled inexorably by, it was inevitably difficult for the MINI to seem as novel as it once had. Some critics voiced their opinions that the cars, much as they drove superbly and were very solidly built, had sacrificed some of their own quirkiness to take on more BMW-like characteristics. Nothing illustrated the ruthlessly

CONTINUED ON PAGE 153

TRAINING GROUND FOR RACING DRIVERS

The original Mini, through handling properties that, in the early 1960s, proved an absolute revelation in small-car control, encouraged amateur drivers to venture on to the racing grid. The new MINI continued this tradition with the MINI Challenge UK from 2002 in its home country and the MINI Challenge Deutschland from 2004 in the land of owner BMW.

The races involved fleets of MINI Cooper John Cooper Works cars of near-standard specification providing support races to major motorsport events in both countries. As the cars were identical, and set up so the teams couldn't tamper with the engine or transmission to unfair advantage, all the emphasis was on driver skill. Many drivers would progress up the racing ladder to series such as European Touring Cars or the British GT Championship after cutting their teeth on the MINI Challenge, where the level-playing-field spec of the car meant skill and daring were brought to the fore.

For the new second-generation R56 MINI Challenge kicking off in 2010, the Cooper S JCW cars received a power hike from 195 to 220 brake horsepower. Adjustable shock absorbers were specially set up for racing, the suspension was lowered, and there were racing tires on special 17-inch alloy wheels. The aerodynamic package included a racing front air dam, rear airflow diffuser, and adjustable rear wing spoiler, all conceived to cut lift and aid grip. Inside, a welded-in safety cage embraced the bucket seats with six-point harness and a HANS (head and neck support) active safety system carried over from Formula 1 cars—then a first in what was called a clubsport racer.

BMW even provided every team buying one of the cars to campaign with a specially designed pneumatic hoist to help with rapid tire changes, so even budding pit crews could gain valuable experience that could take them on through the motorsport world as they perfected split-second wheel swaps.

Wheel-to-wheel action was familiar territory to the old Mini; in 2007 it came to the new MINI too with the debut of the crowd-pleasing MINI Challenge support race series. The 220-brake-horsepower cars were supplied to teams which also received special matching pit-lane equipment.

[left] The three-door MINI Paceman was part of the range from 2011 to 2016, and was manufactured alongside the Countryman in Austria, with front- or four-wheel-drive options.

[below] A fleet of MINI Es was used in and around the 2012 Olympic Games in London, part of a worldwide fleet of some 600 cars built for feasibility studies.

[above] A MINI Cooper S hatchback (foreground) and convertible in Los Angeles in 2013; unlike in the 1960s and '70s, the revived marque proved a huge hit across the United States.

[right] With most of London's street plan landmarks already taken for previous limited editions, MINI headed to the north of the city and Highgate for inspiration for the 2012 convertible special.

[top left] The 2010 MINI One D Clubman. The diesel small station wagon had great appeal across the whole of Europe.

[top right] The final refresh for the second-generation MINI range (Cooper Ss shown) came in 2010, by which time it was facing stiff opposition in the cute small car market from the Fiat 500.

[above] In 2013, MINI made a surprise return to the small van market with this steel-paneled rendition of the Clubman; it was only on sale for two years.

corporate ethos more, perhaps, than the buying out of the John Cooper Works company in 2008 from Michael Cooper, giving BMW total control over the hallowed brand name and the way it was applied to the MINI. From then on there could never be anything even vaguely amateurishly British, or charmingly incidental, about how the MINI evolved.

Having said that, rivals had been eyeing up the MINI modus operandi. Fiat's revived 500, in particular, was very much in the mold of the R50 MINI—compact, retro, cute, and customizable, with an Abarth heritage to match the sporting bona fides of Cooper. Buyers looking for something new and trendy for the modern motoring environment forgot the sage advice they'd swallowed for years about the unpredictability of Italian cars, relented, and started buying 500s by the boat-load, with MINI sales suffering. Hence, the MINI brand was spread out to the very different Countryman and, in 2012, its two-door stablemate, the Paceman, which was built alongside it in Graz, Austria, and could be had with two- or four-wheel drive to suit the crossover ambitions suggested by its high-waisted stance. This car, perhaps most of all, took the MINI name beyond the bounds of acceptability for many with knowledge of the whole Mini history, and it was axed in 2016 after an indifferent sales performance.

Just before Christmas 2013, the last of the R56-type hatchback MINIs was driven off the Plant Oxford production line. Yes, the mainstay MINI was about to be replaced yet again, and it was leaving with some uplifting achievements in its wake. A total of 1,041,412 had been produced, a large part of the 2.4 million MINIs sold since 2001. For 2012, the MINI had accounted for 14 percent of all the cars Britain made and also 14 percent of all British cars exported. The British motor industry had been pronounced a spent force many times in the past few decades, but the MINI proved it was anything but.

CHAPTER

MINI RENEWAL AGAIN

OX67LUM

THE THIRD GENERATION OF MINI, WHICH LAUNCHED IN 2014, HAS BROUGHT MORE MANUFACTURE ABROAD, THE FIRST FIVE-DOOR MINI, AND—WITH AN ALL-ELECTRIC MODEL DUE—SOME FUTURE-PROOFING FOR THE LITTLE CAR WITH A BIG HEART.

Almost a decade and a half after the world's best-loved small car enjoyed its spectacular revival, the new MINI remained a thorny subject among diehards. Many opined it was too big, not roomy enough, a bit too elaborate, straying too far beyond the original car's principles; somehow a "pretender." For balance, of course, one should point to the impressive sales figures and how the MINI had meaningfully added to the choice available to small-car buyers. But for a marque that traded so blatantly on the fond memories of the car that preceded it, inspired it, and sold five million copies (perhaps fifty million motorists had owned an Austin, Morris, or Rover Mini at various times), the strident opinions were only to be expected. Moreover, the criticisms almost never took into account the fact that the new MINI had taken a quantum leap in safety (contemplating a high-speed crash in the old one was enough to put you off using it) and was in another league for comfort (rubber suspension notwithstanding, the original Mini didn't suit long journeys, had a jarring ride and an awkward driving position, and offered seats almost calculated to bring on backache). Most of the new MINI's detractors were driving other modern cars, anyway, and not the original rust-riddled Minis that had all failed their Ministry of Transport tests (MoTs) years before . . .

Horseshoe-shaped LED day running lights were incorporated into the edges of the big oval headlamps.

Reason alone, though, couldn't stop the third-generation MINI from coming with a new mix of controversies. As before, it was hard at first to detect that the new car was actually totally new from nose to tail because the overall profile and most of the details adhered faithfully to the look so fastidiously established back in 1997.

The all-new chassis platform of this MINI was called UKL, a title that was a little misleading as it had nothing to do with the United Kingdom but signified *Untere-KLasse*, meaning *lower* or *small class* in German. The modular structure, devised in Munich, with its reworked MacPherson strut

Looks familiar, of course, but the third-generation MINI (this being the Cooper) was built on BMW's all-new UKL platform. A longer wheelbase and wider front and rear track meant the car was bigger, with nearly a third more luggage space.

front and sophisticated multilink rear suspensions, was intended to underpin a wide variety of future BMW vehicles (especially the later BMW 2-Series saloon, convertible, and mini-MPV) and had been designed to be adaptable to front-, rear-, and four-wheel-drive configurations depending on model. UKL1, this new MINI Hatch/Hardtop also codenamed F56, was the very first. There were vague echoes of Alec Issigonis's family of front-drive Mini/1100/1800 cars, with the extension at the top of the range to the rear-wheel-drive Austin 3-Liter. But what this new MINI actually relayed was that this was an entirely German-designed vehicle, with its engineering, development, and intellectual property having no input from the Brits. A highly distinctive rival, the Nissan Juke, was arguably more of a wholly British car, as it had been designed, partly engineered, and built in the island nation, but Nissan rarely resorted to swathing it with Union Jack imagery in the MINI style . . .

The MINI had grown considerably. It was just under 10 centimeters longer on a 2.8-centimeter-longer wheelbase. There has long been a tendency in general for replacement models to become ever-bigger cars, but in the case of the MINI the increase was needed to reshape the nose to meet ever more stringent pedestrian-impact rules. The grille was larger, and the headlights, innovative units with LED daytime running lights around their chrome-rimmed perimeters, were more raked

back. The front overhang was bigger and so were the rear light clusters. The flanks of the car were still clean-lined, but wider track (42 millimeters more at the front, 34 at the back) made the car appear noticeably more rounded. The larger-than-life exterior, though, didn't result in a great deal more cabin space, although 30 percent extra luggage room was provided.

Under the clamshell hood, which like the Mark II version left the headlamps in situ as it lifted, was a brand-new engine, but this time nothing was shared with Peugeot as had been the case in the outgoing car. Here was a brand-new three-cylinder motor owing all its design and manufacture to BMW, and one that would find its way into several cars carrying BMW's logo, including the i8 hybrid-electric supercar, for which it was used as an acceleration booster, a generator, and get-you-home back-up should the lithium-ion batteries be out of charge. For the mainstream MINI Cooper it was a 134-brake-horsepower turbocharged unit at 1,499cc, mated to a standard six-speed manual transmission.

Critics loved the car. *Car* magazine found the burbling new engine had transformed both its performance and character. There was 16 brake horsepower more over the outgoing, nonturbo Cooper, but torque had leapt from 118 to 162 pounds-feet from a mere 1,250 rpm, making the Cooper a proper hot hatchback. "Sixty-two

The new MINI was sold in 110 markets around the world, reached via thirteen weekly shipments like this.

[top left] Press here for action; a bright red start button was a new focal point of the driver interface.

[top right] A Cooper S mixing it with London's traffic throng; with 0 to 62 miles per hour acceleration in 6.8 seconds, there could be no doubting the car's rapid reactions.

[above] In 2013, the arrival of the five-door Hatch made a regular MINI a comfortable proposition as a family car for the first time. Two extra rear doors were carefully incorporated into the familiar MINI package.

appears on the new speedo in 7.9 seconds, a massive 1.1 seconds quicker than before," wrote its road tester. "But it's the in-gear shove you notice the most, that easily accessed big-car surge that makes climbing hills and overtaking easier. So it's a much more usable engine than before, and hugely more frugal too." With its overall fuel consumption of 60-plus miles per gallon, reduced by a fifth over the old car, the magazine questioned why anyone would really want the 113-brake-horsepower three-cylinder turbodiesel option in the Cooper D, especially as it cost £1,150 more than the £15,300 gas car. And that was even before the statistical credibility of many diesel engines was eroded by the Dieselgate scandal. One aspect the tester wasn't enamored with was the gearbox, feeling that the space between the ratios—second gear ran to 70 miles per hour, and third to 100—hampered the engine's sporting eagerness. But the tester loved the steering, the gearchange, and the MINI's "ability to entertain . . . that trademark agility."

There was also a MINI for little old ladies and driving schools. The new MINI One had a mild yet lively 1,191cc version of the engine, its 75 brake horsepower perfectly adequate for the suburban commute. After all, this was a car now offering refinement well up to the standards of the BMW 3-Series, if not in serene absorption of every road surface imperfection then certainly in noise suppression. The quality of the cabin was vastly improved too. Most of its features were instantly familiar to existing owners, although switches were more logically positioned and the speedo had

vacated its position in the central circular display unit to find a new home behind the steering wheel; a BMW-style "infotainment" sat nav/in-car entertainment display was now installed centrally, still with an iDrive manual controller between the seats rather than a more intuitive touchscreen. Every model now had Bluetooth and air conditioning as standard, plus keyless start to use in concert with the new toggle-style starter button, while the options list included such contemporary must-haves as adaptive cruise control, adjustable shock absorbers, LED headlights, and a reversing camera.

The MINI Cooper S, meanwhile, for the first time diverged from the standard Cooper with an entirely different engine, a four-cylinder 1,998cc BMW turbo with more of everything—189 brake horsepower and 206 pounds-feet of torque that could be upped momentarily to 221 pounds-feet when the turbo's Overboost chimed in. Zero to 62 miles per hour now took just 6.8 seconds. A Cooper SD model ran a four-cylinder BMW turbodiesel.

The new cars were revealed in November 2013 and started appearing on the road by March the following year. Just three months afterwards, marque devotees were in for their next culture shock, as the first ever five-door MINI hatchback was revealed: the F55/UKL2 variant, which showed the versatility of the platform with a 15-centimeter increase in length.

[top] Getting the blues: a MINI five-door hatch bodyshell losing its undercoat look in the automated Plant Oxford paint shop . . .

[above] . . . And here is very possibly the same car nearing the end of the manufacturing process, and showing the headlights that remained in place when the hood was lifted.

[left top] It's intriguing to wonder what Issigonis would have made of the sat-nav screen today in the middle of his precious central speedo.

[left middle] Union Jack imagery remained a distinctive MINI feature in the options list . . . even if the car was completely designed in Germany, and the convertible built only in the Netherlands.

[left bottom] Surprisingly good baggage accommodation in the MINI convertible as the boot lid dropped down and the hood lifted up for access.

Cramped rear space had been the bugbear of all new MINIs until now. The five-door suddenly made an elongated MINI hatchback a serious proposition for the small family.

In September 2014 when the five-door Hatch went on sale, one of these cars, in Cooper S spec, became the three millionth MINI to roll out of Cowley's Plant Oxford. On the same day, the two millionth MINI for export—a Hatch in Volcanic orange, apparently—began its lengthy overseas odyssey to Japan, just one of the 110 markets the car was sold in. The United States, incidentally, was the biggest destination territory, followed by the United Kingdom, Germany, China, and France. Eleven double-deck trains carried completed export cars every week to Purfleet, where they boarded thirteen ships for international dispatch.

Having just invested a further £750 million in its British MINI factories on top of the £1 billion already spent on the three MINI manufacturing facilities, BMW could now watch with satisfaction as its 4,500-strong Oxford workforce produced one thousand cars a day, more than three times the rate achieved in 2001. In 2015, for example, 340,000 MINIs found customers, which included the newest iterations introduced that year, the Cooper S John Cooper Works editions and the third-generation of the ever-popular Convertible.

Nonetheless, the human workforce had some help from the 1,500 robots that performed every one of the 6,000 spot welds on each MINI as its 435 panels came together in the Oxford body shop, overseen by 502 Perception Measuring cameras.

CONTINUED ON PAGE 164

THE MINI CLASSIC: LIFE AFTER DEATH

British Leyland may have been regarded by critic and citizen alike as a disastrous undertaking when it went bust and was nationalized in 1975. Yet, at the time, someone in the organization had the foresight to realize the company accounted for the lion's share of Britain's indigenous motoring history. In order to shield this legacy from the harshness of the mid-1970s economy, a bright spark decided to formalize all the company's heirlooms in a Heritage Collection, and when BL Heritage Ltd. was created in 1979 there was then a proper entity to preserve records. Eventually it would become the British Motor Industry Heritage Trust.

This organization went further than merely looking after vehicles and dusty archives—much further. By the mid-1980s it was stepping in to save irreplaceable original tooling from scrappage, and in 1988 it started remanufacture of complete body shells for the MGB roadster so owners could replace the often highly corroded bodes of their cars while transferring over mechanical parts, interiors and, of course, the original chassis numbers.

This proved such a success the scheme was extended to the MGB GT, MG Midget, and Triumph TR6. There proved to be thousands of enthusiastic takers for these brand-new originals, and the enterprise rapidly became rather more than a cottage industry.

The management of British Motor Heritage bought the business from BMW in 2001. Shortly before that the team had undertaken its biggest ever project by acquiring all the tooling for the old Mini, being able to save everything they needed to by having first dabs on the unwanted production line at the Longbridge factory.

The gradual availability of complete new bodyshells for the Mini MkI and MkII and the Clubman/1275GT, has been a huge success, and many thousands have been built and sold even at the hefty price (at the time of writing) of £9,950 for a Mark I shell. Some forty employees at the company's Witney, Oxfordshire, facility hand-assemble the bodies from parts stamped on the genuine original tools. You simply can't get more authentic. The only stipulation, as ever, is that the buyer needs an original car, no matter now dilapidated, to lend its chassis number and legal legitimacy to the new shell.

The virgin shells have allowed some spectacular restorations to take place in enthusiasts' garages the world over. Meanwhile, new entrepreneurs have been motivated to take advantage of this shiny new metalwork to create yet more alternative Minis. The David Brown Automotive Remastered Mini unveiled in 2017, for example, offers ultimate luxury and retro style using a new Heritage shell, a 78-brake-horsepower 1,275cc engine, coachbuilding techniques to remove the body seams and line up the doors for perfect panel fit, standard satellite-navigation and keyless-go systems, central locking, and USB connectivity. The interior and trunk were lined with top-notch leather, while the flawlessly finished paint process took a daunting four weeks. Indeed, 1,400 man-hours went into each one. A bespoke snip at £75,000.

[above] The Remastered by David Brown Mini took 1,600 man-hours to produce, and was a wonderful way to blow £75,000!

[left] Because British Motor Heritage acquired all the original tooling, it can now offer complete Mini shells to give rotten originals a totally new lease on life. Supplied ready-primered but requiring everything to be swapped over from an existing car, a Heritage shell also took on the existing VIN as part of the major transplant.

Such was BMW's zeal to increase sales, the efficiency of Plant Oxford alone wasn't enough. In search of extra capacity, it spotted the opportunity of the mothballed VDL Nedcar factory in the Netherlands, which had once built DAFs, Volvos, and latterly Mitsubishis and Smarts until those manufacturers abandoned the place. From 2014, it started assembling the Mini Hatch, and within a couple of years it was the sole plant in the world producing the MINI Convertible and the Countryman crossover.

The Dutch clearly did an excellent job because BMW then added the UKL-platform BMW X1 to the mix too. Meanwhile, back at Oxford, the third-generation Clubman (F54) came on stream in 2016. The distinctive twin side-opening cargo doors were redrawn with prominent light clusters laid out across them, but the single side Clubdoor had gone, replaced by conventional rear passenger doors. The effect was to produce the largest car ever to sport the MINI badge, at 427 centimeters long, and for this compact yet capacious shooting-brake to make small inroads into sales of competitor models well outside of the MINI's traditional market sector. At around the same time, a four-wheel-drive option was also offered on the Clubman ALL4, and the complexity of the Plant Oxford production mix was intensified by the fact that a few of the old-style (and ill-fated, as they were destined not to be replaced) Coupés and Roadsters were still inching their way along the lines.

The ALL4 system, a default 50/50 split four-wheel-drive system rather than an extension of front-wheel drive, was eventually offered on Cooper, Cooper S, and Cooper D models. It used an electrohydraulic pump to adjust the clutch, sending up

[top left] No other small estate car had the twin cargo-area doors of the MINI Clubman, a cute link back to the Traveller/Countryman of the early 1960s, even though there was something very un-MINI-like about the thickness of the screen pillars.

[top right] The new Clubman of 2014 now came with two full-size rear passenger doors, making this the biggest MINI so far.

Just one hundred of these John Cooper Works Challenge road car editions were built in 2016, with a semi-racing spec and a 153-miles-per-hour top speed.

to 100 percent of the power to the rear wheels if needed. The lack of raised suspension or knobby tires, though, helped disguise the fact that this was one MINI that could easily cope with muddy country tracks or icy Alpine roads.

The days of the 1970s when the people in charge of Mini were happy for years to pass by with almost no changes to the cars were by now but a distant memory. Toward the latter part of the 2010s, new derivatives came thick and fast. In 2016, for instance, the first sale-promoting, limited-edition three-/five-door Hatch, the MINI Seven, was swiftly followed by a John Cooper Works Challenge special for the United Kingdom and then a brand-new Countryman (built exclusively in the Netherlands), which not only spawned a John Cooper Works version but also a gas/electric plug-in hybrid variant capable of an astounding, super-low-emission 135 miles per gallon.

BMW had already decided to go the whole hog and develop an all-electric MINI, although its path to releasing such a car on the retail market has been a long and cautious one. The venture began in 2008 when it built some six hundred engineless glider MINI Coopers in Oxford, which were sent to company HQ in Munich to have their AC Propulsion–designed electric drivetrains installed. A synchronous electric motor in the engine bay turned the front wheels and drew power from a 572-pound lithium-ion battery pack replacing the back seat.

THE ORIGINAL BUZZBOX

By the time this book is published, production electric MINIs will probably be prowling silently along city streets. Tantalizing images have been released of the car with its blocked-off grille (there's no radiator to keep cool), acid-yellow styling accents, and weird-looking wheels, although full details of the electric drivetrain have not yet been vouchsafed.

The fuel-filler flap replaced by a charging jack is just about the only giveaway that this is an electric Mini; that, and the complete absence of the usual whining transmission racket.

BMW likes to keep MINI momentum going all the time, though. And at the April 2018 New York Motor Show it revealed an interim car that previewed the electric version and, for the Mini's sixtieth birthday year, made a direct link to its glory days.

It was a classic Mini very evocatively and nostalgically turned out with a one-off electric drivetrain. The historic look was carried off wonderfully by the hallowed livery of red paintwork with contrasting white roof and hood stripes on the sympathetically restored base car.

Only the unique badging and emblems hinted at the thirty-cell-strong lithium-ion battery pack and unspecified electric motor inside it.

The original four-speed manual transmission was retained, although there was no longer any need for a clutch, and the car weighed 770 kilograms—the same as a standard 1998 Cooper but the weight of the fuel tank and exhaust system replaced by that of the batteries in a big box where the back seat should have been.

It could do, BMW said, 75 miles per hour, and make 65-mile forays before its charge diminished to nothing. Not too impressive for a car of 2018; then again, this fully functioning electric car was based on a vehicle created in the late 1950s . . . and its emissions were zero.

-O [left] On December 1, 2016, workers at Plant Oxford took their hands and tools to the three-millionth MINI, a Clubman John Cooper Works, fifteen years after the first car was made there.

[below] All-wheel drive, called ALL4, was rolled out across most of the MINI lineup in its third incarnation. This is a Clubman Cooper S ALL4, making light work of the nursery slopes.

[above] Countryman is the MINI purpose-designed for the great outdoors, as well as the school run and supermarket swoop, taking MINI into direct competition with other crossover cars, whose market sector had boomed since the mid-2010s.

[right] A fully loaded Countryman John Cooper Works, ideal for switchback roads and the occasional rough backwoods track.

CONTINUED FROM PAGE 165

The completed fleet of MINI Es were then leased to specially selected users in key markets around the world for two six-month periods. Drivers in Los Angeles, New York, Berlin, Munich, London, Oxford, Paris, Beijing, Shenzhen, and Tokyo took part. BMW claimed there was no bigger electric car field test anywhere, and while most were crushed afterwards, the British fleet survived to be used for promotional activities surrounding the 2012 Olympic Games in London. The maximum theoretical range was 156 miles, although 109 miles was the predicted normal availability in around-town driving. A fast-charge system gave total energy replenishment in 3.5 hours. Users had to complete online logs to record their everyday findings, which were fed back to the development team. Most of the drivers who volunteered were male early-adopter gadget freaks in their thirties, but many of them still reported anxieties about limited charge—especially in very cold or very hot weather—as well as range and space inside the car.

Nevertheless, and despite previously dallying with the idea of hydrogen-powered MINIs as its option for alternative fueling, BMW has decided to unleash the electric MINI to the public in 2019 and to manufacture it in Britain. Fittingly, this is exactly sixty years since the original turned the small-car world on its head. Can we expect a similar revolution? As the Mini/MINI has proven, perhaps more than any other individual car model, anything can happen . . .

[top left] This clay model of the Countryman dashboard gives a good idea of the look that designers aimed for.

[top right] Luxury and bling combined in the interior of the all-new MINI Countryman, here in Cooper S form.

[above] The new Countryman arrived in 2016; this one is a Cooper S, but a plug-in petrol/electric hybrid drivetrain was also available.

ABOUT THE AUTHOR

Giles Chapman writes and talks about the motor car, its industry, its history, and its culture. He is the author of some fifty books on a huge spectrum of car-related subjects, including *My Dad Had One of Those* (selling almost 200,000 copies), *Chapman's Car Compendium*, *Britain's Toy Car Wars: Dinky vs Corgi vs Matchbox*, and *Gentleman Heroes: YU 3250, the First Blower Bentley & the Men Who Made it Happen*. He was editor-in-chief for DK of *The Car Book*, *The Classic Car Book*, and *Drive: The Definitive History of Motoring*, acclaimed reference books that have been published in multiple language editions worldwide. Early on in his career, he edited *Classic & Sports Car*, the world's bestselling classic car magazine, and has contributed to countless other publications. He founded the Royal Automobile Club Motoring Book of the Year Awards, and indeed his own work has received several awards and nominations, including most recently the 2018 Pemberton Trophy from the Guild of Motoring Writers. He owned an original Mini 1000 in the 1990s, igniting a love affair with this eager, characterful little car that has culminated in the writing of this book. www.gileschapman.com

PHOTO CREDITS

AUTHOR'S COLLECTION—GILES CHAPMAN LIBRARY: pages 4, 7 (center left; center right; bottom right), 8/9, 10, 11, 13, 14 (bottom), 15, 16 (both), 17 (bottom), 19, 20 (top), 21, 23 (both), 25, 26, 27, 28 (both), 29 (both), 30, 31, 32, 33, 35, 34 (both), 37, 38, 39 (both), 40 (top left and top right), 41 (top), 42, 43 (top left and top right), 44 (bottom right), 45, 49, 51, 52 (top), 63, 64, 65, 66 (top), 67, 69, 70 (top left and bottom right), 71 (bottom), 72, 73 (both), 74 (all three), 75 (both), 76 (both), 77 (top left and top right), 79, 80, 81, 82 (bottom), 83, 84 (top left), 85, 86, 87 (both), 89 (bottom left and top right), 90, 91 (both), 92 (bottom), 93 (top), 95, 96, 97 (top), 98, 99 (both), 100 (both), 101 (top and bottom right), 102, 103 (bottom left, top right and bottom right), 104 (all four), 105 (top), 108, 109, 110 (both), 111, 112, 113 (both), 114 (top), 115 (both), 128, 129 (all three).

BMH/GORDON BRUCE ASSOCIATES: page 163 (bottom).

BMW: page 22 (lower).

CITROEN: pages 14 (top), 17 (top), 20 (bottom).

DAVID BROWN AUTOMOTIVE/NEWSPRESS: page 163 (top).

FIAT: page 43.

MINI: pages 5 (top left and top right), 41 (bottom), 44 (bottom left), 50 (bottom), 52 (bottom), 53, 54, 55 (all three), 56 (all three), 57, 58, 61, 66 (bottom), 70 (top right), 71 (top), 82 (top), 84 (right), 88 (both), 89 (top left), 93 (bottom), 101 (bottom left), 107, 114 (bottom), 116, 117, (both), 119, 120, 121, 122 (all three), 123 (all three), 124, 125 (top), 126 (all three), 127 (both), 131, 132, 133 (both), 134, 135 (all three), 136 (all five), 137 (both), 138 (all three), 139 (all three), 141, 142, 143 (both), 144 (both), 145, 146, 147, 148, 149, 150, 151 (both), 152 (both), 153 (all three), 155, 156, 157, 158, 159 (all three), 160 (both), 161 (all three), 164 (both), 165, 166, 167 (both), 168 (both), 169 (all three).

NEWSPRESS: pages 18, 22 (top), 40 (bottom right), 47, 50 (top), 59, 60, 77 (bottom), 92 (top), 103 (top left), 105 (bottom).

NICK KISCH: pages 5 (top right), 44 (top).

NISSAN: page 97 (bottom).

INDEX

ACKNOWLEDGMENTS

I am grateful to the good offices of MINI UK, in particular media relations manager Chris Overall, and the Newspress media resource for providing images used in this book. The rest of the pictures are from my own collection; over thirty and more years, I've never been able to resist adding great Mini images to my collection—the more unusual the car or situation, the better. I am also grateful to Nick Kisch, who provided the wonderful image of the op-art Mini used on pages 5 and 44.

Appreciative thanks must go to my fellow car writer, the very eminent Richard Bremner, who kindly read the manuscript and highlighted aspects that I had either overlooked or not appreciated fully. Designer David Saddington was enlightening on the development of the new MINI, in which he played a crucial role. Also to James Mann, the excellent car photographer, who first suggested me as the potential author of this book to the publisher, Zack Miller. Working with Zack, Alyssa, Dennis, and the team at Motorbooks has been a delight.

There have been quite a few books on the Mini over the years, although none that take quite the broad automotive and cultural overview of this one. There are a couple I'd suggest any enthusiast has on the bookshelf alongside this one. *The Complete Catalogue of the Mini* by Chris Rees (Herridge & Sons, 2016) is a fantastic resource of comprehensive facts, figures and data on the original Mini, while *Mini: The Definitive History* by Jon Pressnell (Haynes, 2009) is a highly detailed investigation into the car's engineering and development. I'm also a big fan of the work of Jeroen Booij, whose two-volume *Maximum Mini: The Definitive Book of Cars Based on the Original Mini* (Veloce, 2009) contains much to fascinate anyone drawn to quirky Mini creations. Finally, *The Mini Story* by Laurence Pomeroy (Temple Press, 1964) and *Amazing Mini* by Peter Filby (GT Foulis, 1981) are contrasting, interesting, and entertaining accounts from the days when the original Mini was still very much in its prime.

GB SN3 4PE
COOPER
GB GKN447N
COOPER